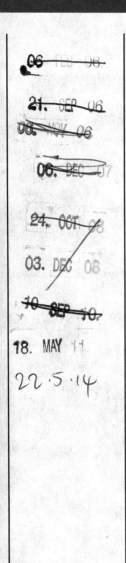

BURN UNIT

BURN UNIT

SAVING LIVES AFTER THE FLAMES

BARBARA RAVAGE

DA CAPO PRESS
A Member of the Perseus Books Group

Copyright © 2004 by Barbara Ravage

Designed by Trish Wilkinson
Set in 11.5-point Adobe Garamond by the Perseus Books Group

Library of Congress Cataloging-in-Publication Data

Ravage, Barbara.
  Burn unit : saving lives after the flames / Barbara Ravage.— 1st Da Capo Press ed.
    p. cm.
  Includes bibliographical references and index.
  ISBN 0-306-81346-7 (hardcover : alk. paper)
  1. Burn care units. 2. Burns and scalds. I. Title. RD96.4.R37 2004
617.1'1—dc22

                                                                    2004003399

First Da Capo Press edition 2004

Published by Da Capo Press
A Member of the Perseus Books Group
http://www.dacapopress.com

Da Capo Press books are available at special discounts for bulk purchases in the U.S. by corporations, institutions, and other organizations. For more information, please contact the Special Markets Department at the Perseus Books Group, 11 Cambridge Center, Cambridge, MA 02142, or call (800) 255-1514 or (617) 252-5298, or e-mail special.markets@perseusbooks.com.

1 2 3 4 5 6 7 8 9—07 06 05 04

To my father, Norman Joseph Kelman, MD
(1914–1997)

Contents

Author's Note

A word about my method. Massachusetts General Hospital and the people who work there were unfailingly generous with time, information, and access, far beyond anything I could have expected. Nonetheless, the protection of patient privacy and confidentiality always took priority, by law as well as my personal ethics. A severe burn is a devastating event for both patient and family. Patients in the burn unit are either unconscious or fully absorbed in the painful process of recovering from their wounds, and they may not be asked to speak about their ordeal while they are in the midst of it. Their families are living through the most anguished time of their lives, and although some consider it acceptable for strangers to intrude upon the anxious and the grieving with cameras, tape recorders, and notebooks, I was unwilling to do so. Instead, I was lucky enough to be introduced to two former burn patients and their families who agreed to share their memories of the experience and to allow burn unit staff to speak to me about their care. I had lengthy interviews with Tom and Nancy Parent, and with the patient I call Dan O'Shea and his parents, here named Jack and Peggy. I spent many hours over the course of more than a year on Bigelow 13, the floor that houses the burn unit at Mass General. I stood alongside nurses at bedside, attended clinic, sat in on informal chats and formal conferences, joined rounds, spoke with attendings, residents, therapists, patient care associates, and other support personnel, and watched them at their work. I witnessed a number of surgeries during a series of hours-long visits to the OR, and toured the Emergency Department at Mass General, the Hyperbaric Medicine Center at Massachusetts Eye and Ear Infirmary, and in the course of a single very long day, every department and floor of the Shriners Hospital for Children in Boston. I had the enormous privilege of spending several hours with John F. Burke, MD, whose contributions to modern burn care are legion. Over the course of two meetings, this raconteur extraordinaire treated me to a minicourse in the history of burns and his not insignificant role in it. I made two visits—once before and once after September 11, 2001—to the Cornell Burn Unit at New York-Presbyterian Hospital, where the few surviving burn victims of the attack on the World Trade Center were cared for. I also attended workshops and educational programs at the annual meeting of the American Burn Association for two years running, where I learned from burn care professionals at institutions throughout the country.

The names, words, and actions of all the burn care professionals used in this book are real; thoughts attributed to them are based on what they told me. The same is true of Tom and Nancy Parent and, except for their names, which I have changed at their request, of the O'Sheas. Other patients whose treatment I describe are composites. In order to comply with the federal Health Insurance Portability and Accountability Act (HIPAA), which governs patient privacy, I have changed the names and identifying details of the patients, in some instances combining case histories in the interest of presenting a comprehensive picture of the course of burn treatment in the most economical way possible. The details of their care, however, accurately reflect what I saw for myself or was told about by people who work with burn victims and survivors.

Burn patients are exposed to a degree that should be seen only by their caregivers and, in certain circumstances, their closest family and friends; their stories are, by and large, unutterably sad. My method reflects the wish to allow a glimpse behind the curtain to those who have never been in a burn unit without breaching the privacy of those who did not or could not give their consent.

Introduction

Ever since Prometheus ignited the first flame, fire has had the power to both help and harm humankind as no other force can do. As a scourge it knows no equal. Medical science may have conquered polio, may be on the way to eradicating malaria, and may in the near or distant future have the tools to stop cancer from taking human life, but fire and the injury it brings will always be with us.

The history of burn treatment is almost as old as burns themselves. The earliest known writings on the subject are found in the Ebers Papyrus (1534 B.C.E.), which prescribes a combination of concoction and incantation to heal wounds inflicted by fire. It outlines a five-day regimen that begins with black mud, progresses through a yeasty dough made with calf dung, a porridge of barley and acacia resin, a paste of boiled beans and beeswax, and ends up with an ointment containing red ochre and copper. Another treatment features a poultice of ram's hair moistened with the milk of a woman who has recently given birth to a male child. The dressing is applied while the healer chants, "Water is in my mouth, a Nile is between my thighs, I have come to extinguish the fire."

Ancient Chinese, Greek, and Roman pharmaceutical texts dropped the magic spells but retained the use of animal, vegetable, and mineral substances, which they claimed promoted rapid healing and minimized scars. These recipes were kept alive in both the Middle East and Europe throughout the Byzantine and Medieval periods by the writings of the seventh-century Greek physician Paul of Aegina and the *Aqrabadhin* (Medical Formulary) of Abu Yusuf Ya'qub ibn Ishaq al-Kindi, the ninth-century polymath from Baghdad. Like those the Egyptians used, most of the treatments were outlandish—bear fat rendered in red wine and rubbed on the skin with roasted angleworms, ashes of dormice, bull bile, rabbit and pigeon droppings, and scores of other ingredients from the catalogue of what has been termed "excremental alchemy." Some make a sort of crazy sense: clays, waxes, and fatty salves are soothing and protective; rubbing burns with lentil meal and honey undoubtedly hurt like hell but would remove dead skin, making the wound more likely to heal than if left to fester beneath burned flesh. Some prefigure treatments used well into the twentieth century, such as the astringent tannin-rich tinctures made from extracts of tea, walnut, and oak leaves. As for the cutting edge, the active ingredient in some of the most effective antimicrobial agents in use today is silver, the antiseptic properties of which were first recognized in ancient Persia and its wound-healing powers mentioned in texts attributed to a fifteenth-century monk and alchemist named Basilius Valentinus.

The focus throughout the ages was on treating the surface of the burn. But whether they were rubbed with a frog dipped in oil, coated with raw egg and vinegar, anointed with pig fat and squill root, or painted with aniline dye, superficial burns healed without scars and deeper ones healed with scars and often crippling contractures. Regardless of how they were treated, people who suffered deep or extensive burns, and even those who were exposed to fire but seemed not to be burned at all, kept on dying. No one could explain why. Then, in a period of amazing activity in many different areas of medical research, the tide turned. Not surprisingly, that period spanned the years between the two great wars of the twentieth century and culminated in 1942, when a deadly blaze leveled a Boston nightclub called the Cocoanut Grove, killing nearly 500 people. The severely burned survivors provided researchers with an opportunity to study burns in all their complexity and improve burn treatment for future generations. More has been learned about burns in the past sixty years, and more lives have been saved as a result, than in all the millennia since the big bang. The key lies in understanding that a burn is an evolving wound; the initial injury is only the beginning of an ongoing disease process that must be stopped or it will end in death.

Heat inflicts a wound with far-reaching effects, some of them paradoxical, others bizarre, many of them not yet fully understood by those who study and treat burns. People who suffer thermal injury are at high risk for hypothermia—abnormally *low* body temperature. The deeper the burn, the *less* painful it is. Early on, burns excite the immune cells and bring about a violent inflammatory response; over time, an exhausted immune system leaves burn victims prey to deadly infections from within and without. People who are in fires often die without any outward sign of having been burned; they suffer respiratory collapse as their airways narrow and close. People who are burned may literally drown as a tidal wave of fluid surges from the blood vessels to flood their lungs. Their hearts may stop, their kidneys shut down, their gastrointestinal tracts become paralyzed. They may swell grotesquely; they may become emaciated. They may starve to death as the wound causes the metabolism to accelerate wildly, turning the body into a raging machine bent on consuming itself. It was only after the Cocoanut Grove fire, when attention shifted from the burned surface to the tempest raging within, that lives began to be saved.

* * *

Over one million Americans are burned every year, more than half of them children. An untold number of burns never come to the attention of a medical caregiver. They are treated with first aid at home, if they are treated at all. Others are tended to in a doctor's office, clinic, or hospital emergency room, and the patients sent home to heal. But each year, an estimated 45,000 people are burned seriously enough to require hospitalization and one tenth that number dies.

It is a common enough story. You read about it in the paper or see it on the evening news. Someone is trapped in a burning building and does not survive the trip to the emergency room. A gas tank explodes in a three-car pileup and some of the occupants burn to death while others are thrown free of the wreckage. A child is accidentally scalded when a pot of boiling water is upset on the kitchen stove. One hundred young people out for a good time are incinerated when fireworks set off an inferno in a rock and roll roadhouse. Fire and heat injury cripple and kill people every day, but unless our own lives are directly affected, few of us know what actually happens to burn victims—both those who die and those who survive.

I began thinking about burns a number of years ago, long before the 2003 Station fire in Rhode Island, even before thousands met fiery deaths on September 11, 2001. The idea of writing a book was sparked by a news bulletin about a building fire and a firefighter trapped in the blaze who was rushed to the nearest burn unit. I had heard such stories many times and had always filtered them through a lifelong New Yorker's large-mesh sieve. But for some reason, that particular story snagged my attention and left me wondering: What is a burn unit? What goes on there? What do they do for a person injured in a fire? What kinds of burns and what kind of care make the difference between life and death? Those questions set off a chain of events that eventually resulted in this book.

The first person I talked to was my father, with whom I shared a penchant for asking open-ended questions. I'm curious about burns, I told him. Do you know anything about them or where I could go to find out? I was surprised when he told me his old friend and medical school roommate, Stanley Levenson, had long been an important figure in the burn world and still taught and did research in the field. He rooted around in one of the endless piles of books and papers that littered his office and dug out a copy of a recent *Harvard Medical Alumni Bulletin*. He suggested I read it and then call Stanley.

Stanley Levenson was a familiar figure at family holiday parties, but I had always lumped him with the rest of the grown-ups. He was just an old friend of my father's; I had no idea what he did for a living. Now an adult making my own living as a medical writer, I still did not know that he was for years the director of the Surgical ICU/Burn Unit at Jacobi Medical Center in the Bronx. I certainly had no idea he had been a young burn research fellow at Boston City Hospital when the first victims of the Cocoanut Grove fire were carried to the Emergency Room that awful night in 1942. Nor did I know until I read the alumni magazine that the Cocoanut Grove fire was one of the worst civilian fire disasters in our nation's history or that it marked a watershed in the medical understanding and treatment of burns. Until I sat in Stanley Levenson's book-lined office at Albert Einstein College of Medicine, what I did know about burns could be written on a matchbook and what I did not know could, and did, fill vast volumes.

As I listened to Stanley give me a brief rundown on the world he has inhabited for more than sixty years, I discovered that the answers to my questions were far more complex, and infinitely more fascinating, than I ever imagined. In the course of that first conversation, I realized that a lengthy quest lay before me. Stanley's office was a good deal tidier than my father's, so he easily found and plucked a book from a shelf: H. N. Harkins, *The Treatment of Burns*, copyright 1943. That would be a good place to start, he told me.

As often happens, life intervened, and it was not until the autumn of 2000 that I found myself talking to Bob Dembicki, the able and energetic nurse manager at New York-Presbyterian Hospital's burn unit, better known as the Cornell Burn Center. After showing me around and telling me about the work they do there, Bob steered me to the American Burn Association, which has since 1967 united the diverse disciplines that make up the burn care community. The next annual meeting of the ABA was in April, he said. I really ought to try to attend. It was my luck that it was to be held in Boston, not far from where I was living at the time.

At the meeting I realized that one of the most remarkable things about modern burn care is its multidisciplinary nature. Other medical specialties rely on a range of health care professionals, but the team approach that characterizes burn treatment is unique. When people in the field speak of the burn care community, they are not simply paying lip service to an ideal; they are living proof of it. This is a broadly collegial group of people, and though not without its hierarchy, it is admirably democratic, in principle and in practice. There are the doctors and nurses, of course, but equally important are the respiratory, physical, and occupational therapists, the dietitians and speech pathologists, the patient care and laboratory technicians, the social workers, psychologists, and pastoral and other counselors with the training and skills needed to reconstruct lives damaged by fire.

I came to understand that the specialized care required by burn victims begins with the paramedics who bring them to the ER and continues for hours, days, weeks, months, and even years. Burn treatment as it has evolved over the past sixty years and as it is practiced today requires a depth of knowledge, experience, and commitment that gives new meaning to the term *intensive care.*

"Burns have a personality of their own," a nurse once told me. "They are visually and emotionally overwhelming." And they present the professionals who treat them and the researchers who study them with challenges unrivaled in the annals of medicine. Mass casualties like those at the World Trade Center haunt us; large numbers of badly burned survivors like those who escaped the Station fire inspire our pity. But in the end, every fire victim is an individual, a singular life to be mourned or saved. Whether it is a crime or an accident, a case of negligence or ignorance, a natural disaster or an act of war, there will always be fires and people will always get burned. There will always be a need to develop better burn treatment and a deeper understanding of how burns kill.

* * *

The first issue of what was then called *The New England Journal of Medicine and Surgery and the Collateral Branches of Science Conducted by a Number of Physicians* carried an article titled "Observations and Experiments on the Treatment of Injuries Occasioned by Fire and Heated Substances." The year was 1812 and the author was Jacob Bigelow, a noted Boston physician who practiced at Massachusetts General Hospital and taught at Harvard Medical School in the first half of the nineteenth century. He was writing in the depths of the dark ages of burn care, when little was known about how to treat burns. It was a time of trial and mostly error, when the only thing that saved victims of severe burns from a lifetime of disfigurement was a painful death. Observing that "the distressing effects of these injuries, when they exist in an excessive degree, are exceeded by few diseases," Bigelow described a state of affairs that persisted well into the twentieth century. "The peculiar appearance of a burnt surface has commonly been supposed to require a peculiar treatment; and many practitioners, instead of resorting to the general remedies of inflammation, have placed their reliance on the supposed powers of a specific remedy. In this way different and opposite modes of treatment have been adopted, whose apparent success or failure at different times has occasioned considerable disputes respecting their comparative efficacy. After a variety of trials have been made, and a multiplicity of cases detailed, the practice still remains ambiguous and undecided; and methods of treatment diametrically opposite at the present day, enlist nearly an equal number of advocates."

There is a building at Massachusetts General Hospital named for Jacob Bigelow and his son, Henry Jacob. On the thirteenth floor of that building is one of the finest burn centers in the United States. Although the geography is undoubtedly coincidental, the history of modern burn treatment is intricately tied to Mass General. This book is an account of how burns are treated today, told against the backdrop of that history and an understanding of the complex physiology of the burn wound and the body and mind it so profoundly affects. There is no better way to tell that story than by focusing on the remarkable team that rescues lives devastated by fire on Bigelow 13.

The history of burn treatment is one of chance discoveries, fruitful collaborations, and key people being in the right place at the right time. State-of-the-art burn treatment is the result of that history and those scientific discoveries. It is also the result of a unique community of healers. Across the entire spectrum of burn care can be found people who are exemplars of selfless, compassionate dedication to the patients and families they serve. So in a larger sense, this book is a tribute to the entire burn care community, past and present. It is a tale of hope and heroism, of lives lost and lives saved, of medicine at its very best. At a time when many have lost faith in the medical profession, it offers a glimpse at a different world: a Level I burn center, where nothing is more important than saving lives, and the history and science behind the everyday drama that is played out there. This is a book about good healers and good medicine, of selfless crusaders, teamwork, and passionate, patient-centered care.

A Life on Fire

Sometime in the early hours of the morning, the narrow entrance to Dan O'Shea's basement studio apartment began to fill with smoke. Most likely a cigarette had been left smoldering in an overflowing ashtray. No one knows for sure. Nor does anyone know when the acrid smoke began to seep into Dan's consciousness as he lay on the other side of the door, sleeping off another alcoholic Friday night. At some point, a smoke alarm went off, alerting other residents of the South Boston townhouse, one of whom dialed 911 before fleeing the building.

Perhaps it was the alarm rather than the smoke that finally penetrated Dan's stupor.

Stumbling in the darkness, he did exactly what you are not supposed to do. As he opened the door leading to the windowless vestibule, a fireball engulfed him, flinging him backward into the room, his face buried in the futon mattress, his back exposed to the flames.

That is how the firefighters found him.

They got to the scene quickly, but it was some time before they were able to get inside. The street entrance to Dan's apartment was a wall of flames, the windows were covered by metal gates. By the time they reached him, he was unconscious and barely breathing.

An ambulance was dispatched at 4:39 A.M. and pulled up outside the townhouse four minutes later. By 4:47, its sirens were blaring as it sped through the darkened streets toward Massachusetts General Hospital.

Inside the ambulance, the paramedics wasted no time starting the life support ABCs: airway, breathing, circulation. When Dan was found, his breathing was agonal. That is medic-speak for the last gasps of a dying person. An oxygen mask had been clamped over his face the moment he was rescued, but his chest was not moving. Experience told them that anyone found unconscious in a smoky fire needs more than a face mask to get oxygen to his lungs. Using a laryngoscope to peer down Dan's throat, one of the paramedics could see that the soot-blackened airway had begun to swell. He inserted a plastic tube into Dan's mouth and down his windpipe to a level below his Adam's apple. The half-inch diameter endotracheal tube would prevent further swelling from closing his throat entirely and suffocating him. With the ET tube in place, the medic started pumping 100 percent oxygen with a hand-operated air mask bag unit (AMBU). By mimicking the rhythm and force of normal breathing with the AMBU, he could keep Dan's lungs operating until they reached the hospital and a mechanical ventilator could take over.

At the same time, the other paramedic hung a bag of saline solution on a hook in the ceiling and started an IV line in the bend of Dan's left elbow. The salt water flowing into his veins was all that stood between him and fatal burn shock.

* * *

In a leafy suburb twenty miles away, Jack O'Shea was up early, as usual. He flipped on the radio at his bedside and got the WBZ time check: 4:55. Then came the news bulletin: A fire in South Boston . . . a single victim, male, rescued and taken to Mass General.

Jack bounded out of bed and headed downstairs. The newscaster had given the East Fifth Street address, but he could never remember the exact house number where his youngest son lived. He figured he should check the book Peggy kept by the kitchen phone before getting her all worried for no reason. He was halfway down the stairs when the phone rang. It was the building owner calling to say there had been a fire and Dan was on his way to the hospital.

Awakening to every parent's nightmare, Peggy O'Shea threw on some clothes and finger-combed her short, blond hair. By the time she got downstairs, Jack was already pulling the car out of the garage. She slid in beside her husband, mouthing a silent prayer as he backed out of the driveway. In a few hours, the roads would be filled with Saturday shoppers, but in the predawn quiet they were able to cover the distance in less than a half hour.

CHAPTER ONE

Saturday Night at the Cocoanut Grove

On November 28, 1942, the Cocoanut Grove, a nightclub at the edge of the theater district south of Boston Common, was packed to the rafters. It was the Saturday night after Thanksgiving, and wartime Boston was filled with visitors—sailors on shore leave, young men in other branches of the military having one last dance before going overseas, and an influx of football fans who had just seen Holy Cross humble the local favorite, number one–ranked Boston College. The crowd would have been even larger if the Boston College victory party had not been canceled in view of the 55–12 rout at Fenway Park.

The official capacity of the two-level club was 600, but the popularity of the nightspot combined with the management's general laxity about safety regulations meant that the customers kept pouring in through the single set of revolving doors fronting on Piedmont Street. It was later estimated that more than 1,000 people were in the club that night. The older, swankier clientele chose the main floor venue, where a floor show, complete with a full orchestra and chorus line, was scheduled to begin shortly after 10:00 P.M. Also on street level was a cocktail lounge, which could be entered through a separate entrance around the corner on Broadway. College students, servicemen and their dates, and young couples out for a good time took the narrow enclosed stairway just inside the main entrance down to the Melody Lounge, a smaller piano bar in the club's basement.

As its name suggests, the Cocoanut Grove was decorated in a tropical motif. It featured a lot of bamboo and rattan, crepe paper and tinsel. Upstairs in the main dining room, seven pillars were decked out as artificial palm trees with papier-mâché coconuts. The ceiling was tented with dark blue satin to suggest the night sky. The fabric covered a huge glass skylight that could be rolled open on warm evenings, but on that cold November night the skylight was locked and hidden by the swathing. Fabric draping on the walls obscured several fire exits. The Polynesian theme extended downstairs in the Melody Lounge, as did the management's tendency to hide and lock exits in a misguided attempt to keep customers from skipping out on the check. Once again, blue satin fabric stretched across the ceiling. Fish netting and driftwood adorned the walls, which were paved with bamboo strips. In the corners, fake palm fronds overhung the banquettes. Small spotlights hidden among the palms provided what little light there was in the cheesily romantic setting.

That night, an amorous customer in a corner booth removed one of the spotlight bulbs in an effort to shed less light on him and his date. A little after 10:00 P.M., the bartender dispatched a busboy to replace it. Unable to see the socket in the darkened corner, the young man lit a match, which he held in one hand while he screwed the new bulb in place.

Partygoers at an adjoining table thought they noticed a spark in the nearby palm tree but went back to their merrymaking when they saw nothing more. Imperceptibly at first, that spark ignited an area of the ceiling drapery and then tore its way across the room.

In an instant, fire engulfed the Melody Lounge, despite frantic but feeble attempts by the bartender and his assistant to douse the fire with pitchers of water. The cloth and paper decor fed the flames, which quickly consumed the oxygen in the windowless space and replaced it with smoke. Electric power went out almost immediately, plunging the already dim room into total darkness. Panicked patrons groped their way to the stairwell, the only exit any of them knew about, and that narrow escape route quickly became a deathtrap. A mere handful made it up the stairs, a fireball roaring after them in search of oxygen to feed on. As for the rest, those who were not trampled to death succumbed to the intense heat or suffocated.

It was all over in the Melody Lounge within three minutes, but it took awhile for people in the larger upstairs club to realize they were in danger. Most were waiting for the slightly delayed floor show to begin, and were perhaps distracted by the presence of western film star Buck Jones in the audience. Some may have heard a commotion in the lobby, but they probably chalked it up to a new influx of partygoers. Then someone smelled smoke and yelled, "Fire!" The guests fled as the blaze erupted into the room. There was only one way out of the main dining room: across the lobby and through the revolving door out to Piedmont Street. A double door onto Shawmut Street, behind the club, was concealed and locked. Two other doors out to Shawmut—the stage door and one leading to the dressing room used by the dancers—were locked as well. Musicians, club employees, and customers rushed to the main entrance. They had no choice even though they were running into the black smoke and flames. As they surged across the lobby and into the narrow foyer, they encountered a crush at the main entrance. Bodies had already jammed the revolving doors, trapping many inside the smoke-filled enclosure, inches away from safety. Firefighters who raced to the scene found close to 200 bodies in the area directly behind the door.

Habitués of the upstairs cocktail lounge and a few stragglers from the main dining room made for the Broadway entrance, but the stampede quickly blocked that single door. It opened inward from a narrow vestibule, which in turn gave onto Broadway, a narrow side street despite its name. Few escaped unharmed, and rescue workers found another mound of maybe 100 bodies just inside that door.

The Boston Fire Department was on the scene almost immediately. One truck was already in the area, having been called to a small car fire just down the street. Despite the men and equipment brought by five alarms, there was little anyone could do but stack bodies on the sidewalk to clear a pathway into the flaming structure. The fire was brought under control in less than an hour, but it was too late to save the building or anyone trapped inside. By the time firefighters entered the ruins, they found only dead bodies. Those who escaped did so in the first minutes, some through the revolving door before it was blocked by fallen bodies, most of the others through the double doors out to Shawmut Street, which were found and forced open just in the nick of time. There was more than one story of heroic patrons who returned to the burning building to rescue others. Some survived; others gave their lives in the effort. Among the dead was cowboy star Buck Jones. Among those who escaped was the busboy whose carelessness was believed to have started it all.

An estimated 300 died at the scene and many more were severely burned. In the end, the death toll stood at 492 (some accounts put the figure at 491), matching the number dead in the San Francisco fire of 1906 and making it the worst in Boston's history. The fire resulted from a fatal mixture of overcrowding, flammable decor, inadequate lighting, and unmarked exits; the absence of fire extinguishers, alarms, and smoke sensors; the lack of a fire or disaster plan for employees; and predictable panic among the crowd. The owner would ultimately pay for his negligence and deliberate violations with jail time, but hundreds of patrons paid with their lives.

Many who got out alive were horribly burned, their clothes and hair aflame as they hit the street. Others emerged choking from the thick black smoke, but many others appeared only dazed. While the bodies of the dead and severely burned lined the sidewalk for several blocks, the walking wounded crowded the area outside the club, some crying in pain or from the horror of it all. Wives searched in vain for their husbands, young men and women called out looking for their dates. Emergency rescue workers, many of them volunteers who had rushed to the scene, tried to impose some order. A makeshift morgue was set up in a nearby parking garage for the obviously dead, but many corpses ended up at local hospitals along with the dying and the gravely injured.

The first casualties were sent to Boston City Hospital, the nearest and largest facility. The fleet of ambulances was supplemented by taxis and private cars commandeered for emergency transport, as well as Railway Express trucks (the precursor of today's express delivery services). According to one account, victims arrived at the hospital at the rate of one every eleven seconds, and approximately 300 were dead on arrival. In the end, Boston City got 131 living patients. Massachusetts General Hospital received 114, but 75 of those were DOA or died shortly afterward. The remaining thirty-nine patients were admitted for treatment. About thirty other patients were scattered among ten other hospitals in Boston and neighboring towns, none as well equipped as Boston City and Mass General to deal with the kinds of injuries seen that night.

* * *

Ironically, the fire could not have happened at a better time. In the weeks after Pearl Harbor, the U.S. Office of Scientific Research and Development, godfather of the Manhattan Project and overseer of the nation's scientific and technological agenda, had awarded grants to two groups of Harvard Medical School researchers to launch a crash study of burns in anticipation of massive battlefield casualties in the newly declared war. Burn research units were established at both Boston City and Massachusetts General hospitals, staffed by the most brilliant medical minds of the time.

The burn research units at Mass General and Boston City were already up and running the night of the Cocoanut Grove fire. The town boasted three world-class medical schools—Harvard, Tufts, and Boston University—and several nursing schools, all of which used the two hospitals as their training grounds. Both hospitals had disaster plans in place and a stockpile of bandages and other supplies, and the city had recently conducted civil defense drills as part of wartime disaster readiness. As soon as the enormity of the situation became apparent, all staff members at both hospitals were ordered to report for duty. A radio bulletin brought in more doctors and nurses, as well as medical and nursing students within a thirty-mile radius of Boston. That night, the navy deployed 200 members of its medical corps, the Red Cross sent in 500 workers, and an equal number came from the Nurses Aides Corps. In the succeeding days, hundreds more nurses arrived to relieve those exhausted by the demands of caring for so many critically ill patients. As Oliver Cope, who headed the federally funded burn project at Mass General, wrote, "Had such a catastrophe taken place before Pearl Harbor, the hospital would have been swamped. As it was, the injured found the staff prepared, for the war had made us catastrophe-minded."

Before the fiery attack on Pearl Harbor, burn treatment was just emerging from the dark ages, spurred by research in the years following World War I, when the toll taken by trauma, burns, and infected wounds had shocked everyone to attention. Physicians were beginning to get the idea that there was a lot more to burns than met the eye and a lot more to learn about how to treat them.

When the Cocoanut Grove went up in flames in November 1942, the 171 victims who made it alive to Boston City and Mass General hospitals became subjects in the most comprehensive clinical trial in the annals of burn treatment. The patients exhibited every imaginable burn complication, making it possible to study this most complex of injuries in great detail. New ideas were tried, old methods discarded, and the agenda for burn research was set for the next quarter century.

It would be impossible in the space of a single chapter to recount every aspect of the treatment the patients received and the depth and breadth of the knowledge gained, but three issues emerged as central to saving the lives of the victims of that fire and of all the fires that have come after it. It makes sense, therefore, to focus on what was learned about respiratory injury, shock, and the surface treatment of the burn wound, as well as the ways in which these three apparently distinct matters are intertwined.

* * *

At Boston City Hospital, the first inkling anyone had about the fire was when three men with burns on their hands and faces walked into the emergency room, having made it there on their own. Within minutes, the stretchers began arriving and the trickle became a deluge. "The examining rooms, corridors and waiting rooms were literally filled with dead and dying and with a large number of living who had burns of varying extent and severity." The entire staff was mobilized, operating rooms were prepared for action, and wards cleared to accommodate the expected flood of fire victims.

Stanley Levenson was a young burn fellow at Boston City in November 1942. A recent graduate of Harvard Medical School, he had come to the attention of Charles Lund, the man in charge of the burn research project at BCH. While an undergraduate at Harvard, Levenson had done research on the role of vitamin C in wound healing. In medical school, he had worked on the development of the oxygen mask that would be used by fighter pilots in the legendary Doolittle raid on Tokyo in April 1942. He was a surgical intern at Beth Israel Hospital at the time Lund was getting things off the ground, so he was a bit late in joining the team. He finished his internship and reported for duty at BCH on November 23, 1942. But he was not on duty the night of November 28. He was in bed with a bad stomachache.

Sixty years later, Dr. Levenson remembered that time in vivid detail. "I had started on Monday, and on Tuesday or Wednesday, I came down with acute gastroenteritis, probably from some of the cafeteria food. I was in bed, admitted to the hospital, on the Harvard medical service. I was feeling better, but I was still hospitalized. As soon as Dr. Lund heard about the fire, he called me and said, 'Look, we're going to have an emergency. You better get up and come down here to the emergency room.' Well, I did of course get up and dressed right away."

Like many large municipal hospitals, Boston City was a complex of buildings connected by a maze of underground tunnels. Levenson followed his nose through the subterranean passageways to the Emergency Department. "I walked through the tunnel, and the smell of burned flesh and smoke was almost overwhelming."

As he approached the ER, he could see victims arriving in droves. "The first job was deciding who was living and who was dead. The majority came in dead on arrival. The patients who were thought to be dead were just laid to one side along the corridor and the others immediately went to the emergency room." Observers have described the scene as revolting, with bodies, many of them in formal evening dress, stacked against the wall like firewood. And the smell. Every account mentions the smell, the memory of which lingers more than a half century later.

The ER was a cavernous room bustling with activity. It took about three hours to separate the living from the dead. It was a daunting task, and sometimes mistakes were made. Levenson recalled "a fire victim who happened to be a surgical resident at City Hospital. He'd been off duty and was at the Cocoanut Grove that night. He was put in with the dead, but fortunately one of the orderlies recognized him and also recognized that he was still breathing. He was pulled out immediately and rushed to the OR for an emergency tracheotomy." The surgery to open his windpipe came too late, however; the young doctor died several days later.

Throughout that first night and over the following weeks and months, all hands were needed. That included not only doctors and nurses, interns and residents, medical and nursing school students, but every aide, orderly, technician, janitor, and elevator operator in the 1,500-bed facility.

Levenson and another young doctor named Charles Davidson were assigned to work under Maxwell Finland, an infectious disease specialist who was chief of the Harvard medical services at Boston City. The two trailed after Dr. Finland as he made rounds, examining 131 patients in thirty-one wards spread out across eight different buildings in the hospital, a four- to five-mile trek from start to finish. Except they never finished. They did rounds nonstop, twenty-four hours a day for the first three or four days. "By the time we finished one round, it was time to go around again," Levenson said. Somehow they managed to see each patient at least once a day, and the sickest twice a day or more.

At both hospitals, the overwhelming majority of victims brought in from the fire suffered from respiratory distress—107 of the 131 admitted alive to BCH, and 36 of the 39 living patients at MGH. Some, including those who were dead on arrival or shortly thereafter, had only minor burns and some had no burns at all. Doctors knew that the blue-faced corpses had suffocated, a common outcome of fire in an enclosed space or burial under a mound of bodies. They recognized in the bright, healthy-looking, cherry-red faces of some victims the telltale sign of carbon monoxide poisoning. The sight of others coughing up frothy sputum, some of it pink with blood, reminded the older doctors of soldiers who had been gassed in World War I or the lung damage seen in victims of the Cleveland Clinic fire of 1929, in which burning nitrocellulose X-ray film stored in the hospital archives gave off noxious fumes. But scores of others who showed none of these signs were dying in the midst of the frantic triage or within a day or two of arrival at the hospitals, their respiratory systems shutting down for reasons that were not yet clear.

Max Finland and his two young assistants were charged with monitoring the respiratory status of the patients, providing them an unparalleled opportunity to observe and ultimately solve the mystery of fire-related breathing problems. It was in this rich clinical environment that Finland, Davidson, and Levenson conducted their classic study of inhalation injury, published later in the journal *Medicine* as "Clinical and Therapeutic Aspects of the Conflagration Injuries to the Respiratory Tract Sustained by Victims of the Cocoanut Grove Disaster."

Without the advanced imaging techniques and laboratory analyses available today, the Finland team was limited to taking X rays and using their stethoscopes to listen for an array of noises—wheezing, stridor, and chest sounds called rales, which are colorfully characterized as musical, crepitant, sonorous, or bubbling—that hinted at what was happening inside. Davidson remembered that "we spent hours putting X rays up and trying to figure out what was going on. In addition to the severe burns of the upper respiratory tract, there was something going on lower down."

They saw the definitive picture when they studied the blasted windpipes and lungs of the dead after autopsy. There was the expected damage to the nose and throat resulting from breathing superheated air, but they were mystified by the degree and type of destruction in the lower airway—below the vocal cords and in the lungs. The full length of the airways was swollen, the linings of the passageways stripped raw. They saw evidence of bleeding, wholesale collapse and destruction of lung tissue, branches of the bronchial tree blown up like balloons by pockets of air trapped behind clots of debris. All this they described, cataloguing the symptoms, classifying them according to four grades of severity, and developing a highly detailed picture of inhalation injury through their observations.

They were able to document the "what." The challenge was finding out the "why."

Was there some sort of poison gas and, if so, what was it and where did it come from? Levenson described the clever bit of detective work that finally answered that question at the same time as it explained why some people had life-threatening respiratory complications, others much less severe damage, and a few none at all.

"Dr. Finland asked Charlie and me to get the patients to tell us exactly where they were when the fire broke out and what they did." They questioned anyone who could talk and even asked about their companions who had died. They compiled data on seventy-two people, including where they were at the moment they first became aware of the fire and their route of attempted escape; whether they recalled being exposed to heat and flame, and if they had inhaled some or a lot of smoke; whether they lost consciousness at the scene and their state of consciousness on arrival at the hospital; how large their surface burns were; and the severity of their respiratory involvement. Levenson and Davidson put it all in a chart and an accompanying schematic diagram showing the floor plan of the club and the location of each of the seventy-two victims, which together tell the story of that horrible night in dramatic detail. The diagram is particularly chilling: Each victim starts off as a numbered rectangle, the route of escape a line with an arrow. Numerous lines show a mad dash toward an exit that ends in unconsciousness a short distance away, the rectangle now a circle, half of it blacked out. Some of the paths double back on themselves, in several cases after it appears the person had reached safety and then returned to rescue others. Sometimes two lines move in parallel and then one of them stops while the other continues. The locked doors and other obstacles are starkly visible, as is the small number of rectangles—thirteen—in the Melody Lounge, where it all began.

As they analyzed the data, it became clear to Finland, Davidson, and Levenson that people with the most severe respiratory symptoms were fairly evenly distributed throughout the club. There appeared to be no single location where the worst respiratory damage was concentrated, arguing against a focal area of poison gas. Ruling out nitrous fumes and phosgene, the two leading suspects in the "poison gas" theory, they concluded, "The respiratory irritant was probably part of the hot air, fumes, and particles which resulted directly from the burning of the contents of the various rooms and rapidly permeated throughout all parts of the building." The culprit was smoke and all it contained: carbon monoxide and dioxide, hydrogen cyanide, formaldehyde, hydrochloric acid, ammonia, sulfur dioxide, and numerous other noxious by-products of the combustion of the wood, paper, plastic, and textiles that fueled the inferno. Everyone inside the club was forced to breathe the smoke; what made the difference was whether an individual lost consciousness and therefore was exposed to it for a longer time.

Thus they arrived at what they called "an alternative and simpler explanation. . . . Those victims who inhaled sufficient amounts of these gases [present in smoke] to render them unconscious, and those who may have lost consciousness from other causes before they reached the open air, inhaled the largest amounts of these hot fumes and particles and, therefore, sustained the severest respiratory tract damage." It may seem obvious today that the severity of inhalation injury is directly related to the length of exposure to smoke, but it was a revolutionary discovery at the time.

Among those who did not lose consciousness and suffered little or no respiratory injury were survivors who covered their faces with damp cloths, which both filtered and cooled the smoke. Levenson recalled the story of a sailor who got himself and his girlfriend out of the upstairs club relatively unscathed, thanks to what he calls basic training. "The sailor knew he should get wet cloths over their noses and mouths, so he grabbed some napkins, but he didn't have access to water. So he used what he had on hand . . . he urinated on the napkins, and that saved their lives."

* * *

The study of shock was a high priority for a nation at war, since it was a risk in most battlefield casualties, especially when large volumes of blood are lost. The fact that shock could occur in burn patients, in the absence of any bleeding at all, had been known for over a decade, but no one was sure why it was happening or how to stop it. In 1930, a Yale physician named Frank P. Underhill was the first to describe burn shock, which he had observed while treating the victims of the Rialto Theater fire in New Haven in 1921. Until that time it was believed that burn wounds contain toxins and that was what killed burn victims. Underhill contended that the cause of death was instead a type of shock particular to burns that resulted from the loss of huge volumes of plasma.

Plasma is the liquid part of blood, a rich soup that contains pretty much everything except the oxygen-carrying red cells, the infection-fighting white cells, and the clot-forming platelets that float in it. One of its ingredients is albumin, a protein that under normal conditions keeps plasma where it belongs, inside the blood vessels. A burn is not a normal condition, however. One of the first things that happens following a burn is that large amounts of plasma pour out of the circulation. As the volume of circulating liquid drops, blood pressure plummets, throwing the body into shock.

If a patient is bleeding, it makes sense to stanch the flow and transfuse blood. When only plasma is lost, what remains is a sludge of red blood cells the heart labors vainly to move around the body. Until the time of Cocoanut Grove, saline solution was as close as anyone had come to approximating the composition of plasma. As for stopping the leak, no one had any idea how to do it. Pouring salt water into the circulatory system might dilute the sludge and give the heart something to pump around the body, but without albumin to keep it in place, it did not stay there for long. Plasma, although limited in availability, seemed a promising alternative, particularly since it is compatible with all blood types, making it a more versatile transfusion fluid than whole blood.

The study of blood and blood products was an important initiative of Boston City's Thorndike Memorial Laboratory, one of the nation's most prestigious research facilities. The lab was the repository of a small store of frozen plasma, every drop of which was needed for the Cocoanut Grove victims. Before hooking up with Max Finland and Stan Levenson, Charlie Davidson had been doing research at the Thorndike lab. Years later he recalled the night of November 28, 1942. He had been out on the town with friends when he heard the sirens and rushed back to the hospital to see if he was needed. As he dashed down the basement corridor, he saw F. H. Laskey Taylor, the head of the Thorndike biochemistry lab, rounding up medical students and anyone else he could find to shake containers of plasma that had been frozen for storage. "It takes a lot of heat to thaw frozen plasma. Somebody hadn't thought of that little problem."

As promising as it seemed at the time, plasma is hardly a panacea for victims of burn shock. Collection, processing, and storage of plasma are neither simple nor inexpensive matters. Bacterial and viral contamination are serious hazards. Sterile solutions that can be produced in quantity from water and minerals are considerably more practical, especially in view of the vast quantities burn patients need.

That Saturday night, the most seriously burned patients were given plasma, administered intravenously and in some cases diluted with saline solution to stretch the limited supply. In the early hours of Sunday morning, Harvard Medical School sent over a small amount of human albumin, which was doled out carefully to those in the deepest shock. How much fluid each patient received was determined by a rough estimate of the surface area of the burn—the larger the burn, the more fluid was given—but some patients with relatively small burns were also developing shock, and they seemed to need much more fluid to reverse it. What all those patients had in common was smoke inhalation. We now know what they did not know back then: Damage to the respiratory tree, which represents a huge surface area, must be taken into account when estimating burn size as a guide for fluid replacement.

In the end, 39 of the 131 patients treated at Boston City died. The respiratory cases died early, the worst of them in the first two or three days. A similar scene was playing itself out across town at Massachusetts General Hospital, though on a smaller scale. Of the thirty-nine patients who survived long enough to be admitted to the hospital, only three did not have some degree of respiratory injury. Seven patients died within the first three days, all of them from respiratory complications.

At MGH, all patients with frank or impending shock were given intravenous fluid within fifteen minutes of arrival at the hospital. They received saline or glucose solution until the supply of frozen plasma from the hospital's blood bank had been thawed in pans filled with warm water. Without trying to figure out how much fluid had been lost, doctors administered 1,000 cc of fluid (half plasma, half saline) over the course of twenty-four hours, for every 10 percent of the surface area of the body burned. This is a paltry amount compared to the liters and liters of fluid typically given today, but they were afraid that giving more would result in pulmonary edema—waterlogged lungs—especially since so many patients were in respiratory distress.

In their account of the Cocoanut Grove experience, Finland, Davidson, and Levenson observed that "the type of personnel, the number of cases, the facilities and arrangements for their care, the methods of approach and the details of almost every phase of their management were decidedly different at the two institutions." Perhaps the greatest difference and the one that has the most historical significance can be seen in how the surface burns were treated.

* * *

It had long been the practice to seal burns immediately, not only to lessen pain but also to prevent fluid from gushing out and infection from rushing in through the open wounds. Sealing was thought to "fix" toxins believed to be present in burn wounds, keeping them from poisoning the rest of the body while allowing new skin to grow beneath the seal, protected from air and injury. It was a theory riddled with error. That error was responsible for the fact that the putrid smell of decaying flesh went along with all but the most minor burns as bacteria thrived and multiplied beneath the seal in exactly the sort of moist, dark environment the worst sort of bacteria like best.

Beginning in the 1920s, the sealant most commonly used was tannic acid, a naturally occurring vegetable extract with a long medicinal history. Ancient Chinese pharmaceutical texts prescribed the use of tannin-rich tea leaves for the relief of pain in burns as long as 5,000 years ago, and it was also a well-known folk remedy in the British isles. Folk healers in many different parts of the globe painted burns with ink and infusions made from oak leaves, another source of tannic acid, or covered them with oak and walnut leaves that had been boiled to soften them and release the tannin. Tannic acid entered the modern armamentarium in 1925, when Edward Clark Davidson, a physician at Henry Ford Hospital in Detroit and no relation to Charles Davidson, described its use for severely burned patients. Davidson's method involved opening and scraping off blisters in a painful procedure called debridement, and then covering the wounds with sterile gauze bandages soaked in tannic acid solution. He prescribed leaving the bandages on for up to twenty-four hours until the damaged tissue had been fully tanned, taking on a leathery texture. Alternatively, the tannic acid could be sprayed on and sometimes patients were bathed in the stuff.

Tannic acid may have sealed the wound, but infection remained nearly ubiquitous. In 1933, a physician at Johns Hopkins named Robert Henry Aldrich came up with the idea of using gentian violet, a coal-tar derivative initially developed as a textile dye but found to have a broad range of medicinal uses. Perhaps best known as a stain used to prepare microscope slides, gentian violet could both seal wounds and kill gram-positive bacteria, of which *Staphylococcus aureus* was the most threatening to burn victims. It did not, however, spare patients the agony of repeated debridement. Aldrich recommended spraying gentian violet on burns and leaving them uncovered while warming the patient beneath a tent of bedsheets fitted with a lightbulb to keep the ambient temperature between 84 and 88 degrees Fahrenheit. The spraying, which stained the bedding as well as the patient a bright purple, was repeated every two hours. Anything that oozed was scraped away. Once the wounds were dry and sealed, spraying was repeated every four to six hours for days and even weeks until healing was complete.

Gentian violet produced a softer seal than did tannic acid, an effect that was considered desirable, and it was easier to see if infection was developing beneath it. But develop it did. Infection remained a fact of life for all but the smallest burns; for many patients it was the cause of death. Aldrich started looking for something that would act against a broader spectrum of bacteria. He hit upon a combination of three aniline dyes: gentian violet plus brilliant green and acriflavine, a yellowish dye. The routine was the same: Debride, spray, repeat every few hours, and watch like a hawk for infection until the wound healed. Signs of infection were easy to see: The dye combination tinted wounds a deep purplish black with a golden greenish sheen; infected wounds lost their sheen and turned bright blue.

Aldrich eventually moved to Boston City Hospital and brought the triple dye method with him. Some BCH doctors adopted triple dyes, others stuck with tannic acid. Still others tried a combination of tannic acid and silver nitrate, the first as a sealant, the second for its antibacterial action. Over at Mass General, Oliver Cope had his doubts about them all.

Cope dated his skepticism about the value of sealing wounds in general and doing it with tannic acid in particular to his medical school days, when he watched victims of a 1928 explosion and fire at the Beacon Oil Company in Everett, Massachusetts, being treated at Mass General.

> I remember the old emergency ward had five little operating rooms, and there were some bathrooms and bathtubs for giving patients a bath. All of them were filled; I remember one most vividly. . . . An intern and a nurse were . . . taking off the blebs [blisters], and the nurse was pouring tannic acid on him. I came back a little later to look, and he had died— no fluid. All the attention was riveted on the burn wound. Here were these poor men pouring out fluid from their burned surfaces, and they were dying for lack of general physiologic attention. Well, that stuck in my mind.

It was the first death the young medical student had seen. Nearly sixty years later, Cope was still outraged. "Unforgettable," he said. "For attention to the wrong thing. . . . the surgeon's absorption with the wound."

Proponents of the triple dye method contended that it made for faster healing by fostering the regeneration of epithelial cells, the beginning of new skin. Cope questioned that assumption and decided to put it to a test. Along with a colleague, Bradford Cannon, Cope compared various topical agents, among them tannic acid, triple dye, and a cheap, simple, and much less painful treatment that Sumner Koch and Harvey Stuart Allen had developed at Cook County Hospital in Chicago. Koch and Allen did not scrape the wounds; they covered them with a sheet of gauze smeared with petroleum jelly and wrapped them in a dressing made from a Dagwood sandwich of cotton batting. Greasing the gauze kept it from sticking to the wound; the bulky dressing absorbed fluid from the wound and also kept the patient warm; pressure from the tightly wrapped dressing was thought to reduce swelling; splinting immobilized joints to prevent wounds from reopening as they healed. Cope and Cannon did their tests on graft donor sites, areas where healthy skin had been shaved away to be used to cover wounds. As a control they used untreated and unbandaged donor sites, just the raw wound over which a scab had formed. They found that "what healed most rapidly was the area covered by just plain bloodclot. Vaseline was almost as good. Tannic acid delayed the healing. The triple dyes were the worst of all."

They completed their study early in 1942, and Cope immediately instituted the Koch-Allen method at Mass General. There was no debridement, blisters were left intact; the tightly wrapped dressings were left in place for about a week. By the time of the Cocoanut Grove fire, that was the standard of care at MGH. Recalling their study years later, Dr. Cannon said, "We had proved convincingly that tannic acid and the triple dyes recommended for topical application were injurious to viable epithelium and delayed the healing process. But word sometimes gets around slowly. . . . I remember well the arrival within a few days [after Cocoanut Grove] of a host of 'experts' from Washington who were so impressed by the performance of our staff that they published directives on burn management based on our experience. Rumor has it that quantities of tannic acid were discarded by the armed forces." It was later confirmed that far from protecting the body from burn toxins, tannic acid was itself toxic to the liver. By war's end, tanning was history.

In his report on the surface treatment of the Cocoanut Grove burns, Oliver Cope wrote, "The treatment used on the burns of the skin was unorthodox but the results were gratifying. Its simplicity has much to recommend it when large numbers of burns are encountered in a disaster. . . . The simpler the treatment, the fewer the trained personnel required to administer it." The beauty of the MGH approach, he said, was that it was a one-step procedure that required neither a surgeon nor anesthesia, freeing doctors to treat shock and breathing problems while nurses, aides, and orderlies slapped on greasy gauze, added a tight wrapping, and used rolled-up newspaper for splints to immobilize limbs. Cope insisted that "neither debridement nor cleansing are essential to the good care of burns. Good surgical judgment in their care, as in that of many other diseases, consists of knowing when not to interfere." The fluid in blisters is sterile, he had found, whereas broken skin provides an easy entry for bacteria that normally live on the skin surface. The vigorous scrubbing intended to remove bacteria can injure adjacent healthy skin, enlarging the wound, and is so painful that it requires anesthesia, which increases the risk of shock. In contrast, bland ointment and bandages decrease pain by shielding the open wound from air and touch. In addition to preventing heat loss, the type of dressings they used keep the wound bed moist, which hastens healing.

In contrast to Mass General, where wounds were dressed according to Cope's evidence-based directive, Boston City used an assortment of surface treatments driven by necessity. At the outset, they followed the U.S. Army protocol: cleaning and debridement, then a single application of tannic acid, followed by four applications of tannic acid–silver nitrate thirty minutes apart. But with so many patients needing immediate and continuing attention, this was a cumbersome procedure. Besides, "supplies of silver nitrate were rapidly exhausted and many staff surgeons not connected with the Burn Assignment were called in and were permitted to use the dye treatment with which they were most familiar." As a result, some patients got triple dye only, some got tannic acid–silver nitrate only, and some got a combination. A study out of BCH published a year after the fire compared these treatments, as well as Cope's method, and found much the same thing Cope and Cannon had: Burns smeared with ointment and wrapped with gauze healed faster than those sealed with triple dye or tannic acid.

The problem of infection persisted regardless of what was done to the burn surface, however. It was the dawn of the antibiotic age and no one treating burns really knew how to make the best use of sulfa drugs and the very new silver bullet, penicillin. The usual suspects—*Streptococcus, Staphylococcus, E. coli,* and *Clostridium*—turned up in many of the wounds. Infection continued to be a major threat for patients and a major challenge for researchers. Today burn units are no longer characterized by the stench of rotting flesh, but even with better surgical and pharmacological weapons, infection still rears its ugly head, due to antibiotic-resistant bacteria and the very nature of the burn wound itself.

* * *

After the Cocoanut Grove, Oliver Cope continued to study the problem of burn shock. It was clear to him that seepage through the damaged skin was a minor matter compared to the shift of fluid from the circulation to the tissues surrounding the burn and, mysteriously, in areas far distant from the wound. He was convinced that this fluid shift was responsible for the massive swelling seen in burn patients and was the cause of burn shock. Along with his protégé, Francis Moore, Cope came up with a formula for calculating how much replacement fluid was needed based on how much of the body was burned. Cope was also one of the first to suggest that burns should be removed surgically, and the sooner the better.

Oliver Cope's name is written indelibly in the legacy of the Cocoanut Grove fire, but he did not work alone. He was part of a generation that changed the art and science of burn care forever. Out of the ashes of the Cocoanut Grove came recognition that inhalation injury is a common and potentially fatal consequence of indoor fires, and the need for better strategies to treat it; a deeper understanding of burn shock and efforts to refine fluid replacement as a means to manage it; vital information about the metabolic consequences of severe burns and the central role of nutritional support in wound healing; use of newly developed antibiotics as a weapon in the war against the virulent and deadly infections seen universally in burn wounds; the triumph of simpler and more effective topical treatment over the time-honored, and time-consuming, practice of painting burns with tannic acid and triple-aniline dyes; insight into the immediate and long-term psychological effects of severe burns and the need to support survivors even after their visible wounds have healed; and an appreciation of the role of physical and occupational therapy throughout all phases of recovery. By-products included the development of emergency procedures for treating large numbers of fire victims, a wave of fire prevention efforts and fire safety legislation, the organization of specialized burn treatment units, and a commitment to the multidisciplinary approach to burn care. It also marked the beginning of the downward trend in both burn incidence (due to better public education and prevention) and mortality. Before World War II, a burn covering no more than half the total body surface area was fatal in half of all cases. Today 95 percent of all burn patients can be saved, including some with burns over as much as 95 percent of their body.

CHAPTER TWO

A Riot in the Body

The Cocoanut Grove was neither the first nor the worst major fire of the twentieth century. That distinction goes to the December 1903 fire in Chicago's Iroquois Theater, which killed 602 people when stage lighting set the velvet curtain ablaze and the flames quickly spread to the highly flammable painted scenery. Tragically, the audience was instructed to remain seated while inadequate measures were taken to fight the smoky fire. That cost the lives of nearly one-third of the parents and children who had come to enjoy a holiday matinee.

Every decade between the two events saw a disastrous conflagration that took scores of lives. In 1911, 146 clothing factory workers—mostly women, mostly immigrants—perished in the notorious Triangle Shirtwaist Company fire. Many were trapped in a ninth-floor sweatshop by locked fire exits or were crushed in an effort to escape by way of a narrow stairway that ended at an inward opening door. Others died jumping from windows or falling from a fire escape to nowhere that was torn off the building under their weight. In Ohio, in 1929, a deadly chain of events that began with a steam leak and ended with a fiery explosion claimed 125 lives at the Cleveland Clinic. Nitrogen dioxide, a poison gas released by smoldering nitrocellulose X-ray films stored in the basement, spread through the hospital ventilation system and was responsible for many of the fatalities. It was this fire that struck a chord with doctors investigating inhalation injuries and deaths among the Cocoanut Grove victims. In 1937, the Consolidated School building in New London, Texas, exploded in a fireball when a spark from a faulty switch in the industrial arts classroom ignited fumes from an undetected gas leak. Deaths numbered nearly 300, most of them schoolchildren.

Investigations following each of these tragedies uncovered negligence, often extreme and criminal, and each inspired new fire safety rules and, for a while at least, stepped-up enforcement. The fire safety codes now widely in effect date from the aftermath of the Cocoanut Grove fire. The state of Massachusetts took the lead in banning the use in public buildings of inward-opening doors of any kind and revolving doors unless they are flanked by a minimum of two hinged doors with push bars. The new state fire laws also decreed that the number of exits must be determined based on the square footage of the space and its capacity, exits have to be clearly marked and unobstructed, backup lighting in case of power outage is mandatory, flammable decorations and furnishings are forbidden, and any textiles that are used must be treated with flame retardant. Many other states followed suit. But on July 6, 1944, not even two years after the Boston fire, 168 people perished in Hartford, Connecticut, when a Ringling Bros. and Barnum & Bailey circus tent waterproofed with a paraffin coating was ignited by a discarded cigarette butt.

The second half of the twentieth century had its share of avoidable fire disasters, some of the worst of them in nightspots where the lessons of Cocoanut Grove should have been remembered: the Beverly Hills Supper Club in Southgate, Kentucky, where 164 were killed in 1977; the MGM Grand Casino in Las Vegas, where the death toll was 85 in 1980; and the Happy Land Social Club in New York City, where 87 people died in 1990.

It seems the lessons have to be relearned every decade or so. Part of the problem is that there will always be public officials who can be induced to look the other way. A stampede at a Chicago nightclub in February 2003 that killed twenty-one people was eerily reminiscent of what happened at the Cocoanut Grove. True, the catalyst was pepper spray, not flames, and the number killed was but a fraction of those who lost their lives in the Boston fire, but the overcrowded club, the rush to a single exit with other escape routes blocked in an attempt to prevent customers from sneaking in or out, and the implication that lax enforcement of safety laws played a part were all too familiar. Who could have imagined that it would be quickly followed by a catastrophic nightclub fire that read like a B movie script of the Cocoanut Grove horror?

Three days later, on the night of February 20, the Station, a down-at-the-heels music club in the former mill town of West Warwick, Rhode

Island, was filled far beyond capacity with young people, their numbers swelled by spring break. They had come to hear Great White, a second-tier heavy metal band whose best years were behind them. As is apparently customary on the hard rock touring circuit, the performance featured a pyrotechnical display—fireworks, indoors, in a crowded space. It is hard to imagine anyone thinking that is a good idea, harder still to imagine that it is legal under any circumstances. Hardest of all to imagine is that the club had passed a fire inspection less than two months earlier, despite the absence of a sprinkler system and the presence of lightweight polyurethane packing foam glued to the wall behind the stage and the ceiling above it in a makeshift attempt at soundproofing.

When a shower of fireworks ignited the foam, many in the audience thought it was part of the show. In a grotesque echo of the Cocoanut Grove, a member of the band tried to douse the flames with a bottle of water, since no fire extinguishers were within reach. It was not until the overcrowded space began to fill with dense black smoke that people rushed to the entrance, the path to which was narrowed by a cashier stand. There were other exits, but few knew where they were and those who did could not find the way in the smoky darkness. One was next to the stage, the center of the raging inferno. Another was behind the bar. A third was in a storage area, out of sight and off-limits to patrons. It is not surprising that most people made for the front door, the door they knew about, the door they came in.

It did not take long for that doorway to become clogged with a mass of humanity and for a pushing, shoving throng to build up behind it. Like the crowd trying to escape from the Melody Lounge sixty years earlier, these young people felt the searing heat at their backs as they faced a wall of bodies, some already dead, others writhing and screaming beneath the weight of those piled on top of them. Survivors reported seeing people in flames, their hair and clothes on fire, their flesh seeming to melt from their bodies. Toxic fumes from the foam choked the air and molten globs of the material dropped from the ceiling, searing skin wherever it landed. Bizarrely, the melee was captured on videotape by a local television crew, and the haunting pictures make it clear that panic brings out the worst in people. Many who died were crushed to death or asphyxiated; injured survivors suffered broken bones in addition to their burns.

It took four minutes for the Station to be reduced to smoking rubble. Within hours, the finger-pointing and denials began, but there was more than enough blame to go around, between the band and its manager, the club owners, the fire inspectors who gave the club a passing grade, and the company that supplied the foam—$575 worth of highly combustible "egg-crate" sheeting.

The number of reported deaths hovered just below 100, inching up and down over the first two weeks as the missing were confirmed dead, the hospitalized survivors were identified, and some of them succumbed to their wounds. Injuries ranged from minor to monstrous and numbered close to 200. As was to be expected given the nature of the fire, at least fifty of those hospitalized suffered smoke inhalation, which greatly complicates treatment and increases the risk of dying.

The eighty-one people who escaped death but were badly burned were distributed among eleven hospitals in Rhode Island and Massachusetts in accordance with state and regional disaster plans instituted in response to the terrorist attacks of September 11, 2001. The cornerstones of these plans called for coordinated communications among medical facilities in the region, rapid assessment of the injured, and transport of those in need of specialized care to hospitals equipped to provide it. As was true in the days after Pearl Harbor, provisions made in a time of war ended up saving lives in a very different kind of emergency. Oliver Cope's words ring true; we are once again "catastrophe-minded."

Nineteen victims were sent to Boston, which has three specialized burn units: at Massachusetts General and Brigham and Women's hospitals, and the Shriners Hospital for Children. Thirteen came through the Emergency Department of Mass General just hours after the fire, two more were transferred later. Most were treated at MGH, but for the first time in its history, Shriners accepted adults for treatment, so great was the need for intensive care delivered by experienced burn care professionals. A little over a week after the fire, two of the hospitalized victims died, and a week later a third death was reported, bringing the toll to ninety-nine. Ten weeks after the fire, the hundredth victim died. All had been in critical condition at Mass General, which took the most severely burned patients. The doctors, nurses, and other members of the burn team fought valiantly to save these lives, but big burns compounded by seared lungs, virulent infection, and organ failure won out in the end. Those who survived faced months of hospitalization, with numerous surgeries, intensive nursing, and arduous courses of rehabilitation. Those who were able to return home will be scarred both physically and emotionally, their lives saved but forever altered.

* * *

A burn is, at its most basic, an assault on the skin, by far the largest organ of the human body and its first line of defense against the outside world. The skin of an average adult covers an area between five and seven square feet, or about the size of a king-size bed. In most places it is not much thicker than the sheet on that bed. This vast protective shield helps regulate body temperature. It is one of the portals through which fluids pass into and out of the body and the medium through which vitamin D is synthesized from sunlight. It is the armor that deflects disease-causing organisms and shields the interior from ultraviolet radiation. Within its layers reside hair follicles, sweat and oil glands, a vast network of nerves and tiny blood vessels, and pigmented cells responsible for skin color as well as freckles, birthmarks, moles, and other distinctive markings. It receives sensations ranging from pleasure to pain and broadcasts them to the brain. It is an elastic sheath that covers nearly every inch of the body, allowing the structures beneath it to move. And it is a garment that contributes richly to our sense of ourselves.

When the integrity of the skin is violated by a burn, every one of those features and functions is compromised. The assault can be minor and transient, from momentary contact with a hot cooking utensil or a small drip of candle wax, for example. It can be more gradual but also more extensive, as a sunburn, the pain of which may last a day or two, though potential but invisible damage on the cellular level may take years before emerging as skin cancer. Or it can be rapid, violent, and deadly, as from flame, intense heat, scalding hot liquid, a high-voltage electric shock, or even exposure to a corrosive chemical. When it is violent, the body responds in kind. All hell breaks loose throughout the living organism, all systems respond to the attack, and in that cruel irony so often found in nature, signals are misunderstood or conflict with each other, resulting in responses that do further and deeper harm.

The seriousness of a burn depends on both its depth and extent. Medical science took a long time to figure that out, but even before the age of microscopic pathology, astute observers drew conclusions that were not far off the mark. Ambroise Paré, the sixteenth-century French barber-surgeon who is considered the father of modern surgery, noted that burns of different depths appeared and behaved differently: "if the burn is superficial, pustules and blisters develop unless proper measures are taken and if the burn is deep, scabs or a crust appear which are burnt flesh. The action of the fire causing combustion thickens the skin, rendering it hard and taut, provoking great sufferings, which draw the humours from the immediate neighborhood of the burn and distant parts and convert them into serious aquosities." As an army surgeon in the service of the French King Francis I, Paré had ample opportunity to study battlefield wounds, including the very common but devastating gunpowder burns, during the siege of Turin in 1536. His observations are stunningly consistent with what we now know.

In 1607, a German physician named Guilielmus Fabricius Hildanus (Wilhelm Fabry of Hilden) published the first book devoted entirely to burns, known as *De Combustionibus*. The full title of his sixty-page treatise gives an indication of its scope: *Burns caused by boiling oil, boiling water, molten iron, hot ashes, lightning, or any other hot substance: A treatise in which are clearly described the Difference, Symptoms, Prognosis and Treatment both of the burns themselves and of nearly all of the accidents which cause them.* Dr. Fabry was clearly concerned about prevention as well as treatment.

Paré and others used depth to distinguish between more and less serious burns, but Fabry was the first to divide burns into three degrees, and though more complicated classification systems have since been proposed, those in use today closely echo the ones Fabry described in such detail in *De Combustionibus*. He termed first-degree burns *levissimam* (fairly slight), and described them as red and painful, with swelling and blisters, followed by peeling. Second-degree burns, or *mediocrem* (more serious), were fiery red and painful, with blisters containing yellowish fluid, and a tightening, shriveling, and thickening of the skin. His description of third-degree burns, *insignem* (very serious), is as vivid as it is accurate: "Black skin, or at least turning black, is seen and smells unpleasant if pricked with a scalpel. There is a hard, dry crust and when it falls off a deep, putrid ulcer remains. . . . Not only the skin, but also the flesh, veins, arteries and sinews are burnt, contracted and dried out into an eschar, since their natural moisture is dried up and destroyed by the violence of the fire."

Much of what Fabry understood about burns remains valid today. He correctly observed that the longer the exposure to heat, the more serious the burn. He considered the most dangerous burns to be those that involve the head and face, eyes, groin and genitals. He asserted that treatment decisions should be dictated by the burn's degree. And he knew that complications arise more often in patients with other illnesses: "It is easy to treat burns in healthy, well ordered bodies. On the other hand, it is very difficult in ailing and plethoric bodies where burns easily degenerate into a putrid ulcer, for the pain draws to itself fluids and blood from the body. The affected part then becomes fiery and swollen and various other and severe symptoms follow." He designed braces and splints to be worn during healing to prevent contractures ("retracted sinews and curvature of the joints consequent upon burns") and devised surgical techniques for removing unsightly scars. But when it came to surface treatment, his medicine chest was decidedly old-fashioned, with its bear fat, earthworm juice, and herbs boiled in a broth made from the feet and head of a castrated sheep.

In the 1830s, the French surgeon Guillaume Dupuytren maintained there were six degrees of burns, ranging from superficial injury to the skin surface all the way down through the muscle, bone, and finally "carbonization" extending into the bone marrow. His system was still in use in 1911, when the *Encyclopedia Britannica* called it "now most generally accepted" in its article on burns and scalds.

The familiar terminology of today falls somewhere between Fabry and Dupuytren. A first-degree burn is the most superficial, affecting only the epidermis, the paper-thin outermost layer of the skin, which is little more than a stratified collection of a rather primitive type of cell, squamous epithelium. It has neither blood vessels nor nerves and serves mostly as a waterproof and germ-repellant shield. First-degree burns, of which sunburn is the most common, look bright pink on light-skinned people; more deeply pigmented skin appears darker than normal with red undertones. These burns are quite painful, due to the excited response of the tiny blood vessels and nerve endings that lie just below the epidermis, but there is no open wound and no living structures have been destroyed. They get better on their own within one to seven days as a thin sheet of dead epidermis peels off and is replaced by a new top layer, leaving no scars. They can look and feel alarming, but regardless of how much of the body they cover, they are never life threatening, and it takes little more than time to heal them. Aloe vera or a moisturizing cream helps soothe the pain; preparations containing alcohol should be avoided.

Once a burn gets to the dermis, the second skin layer, the damage and the body's response start to be serious. The dermis is eight or nine times thicker than the epidermis and is much more complicated territory. Elastic and fibrous cells give it structure. Nerves, capillaries, and lymph ducts thread their way through it. A diverse population of specialized immune cells is on constant patrol against invasion. Hair follicles reach down into the dermis, forming channels lined with epithelial cells out of which hair shafts grow. Sweat and oil ducts, also lined with epithelium, reach up from the dermis to pores at the surface.

Second-degree burns penetrate the epidermis and extend into the dermal layer. In the world of burn care, they are referred to as partial-thickness burns and are further distinguished as superficial or deep, depending on how far down into the dermis they go. Superficial dermal burns are moist, bright pink or red, and blistered. They will blanch when pressed and then redden again, which indicates that blood is flowing into and out of the area through intact capillaries. They are the most painful of all burns because the nerve endings in the dermis are exposed to touch and air. They usually heal in a week to ten days with minimal scarring. The appearance of deep dermal burns depends on the heat source—in essence, whether the skin was seared, steamed, or boiled. They may be ivory white or dark red or a mottled combination of the two, but they are dry rather than weepy and do not form blisters. They also do not blanch because the heat has damaged the capillaries and cooked the blood within them. These burns tend to be less painful than the superficial ones because of the significant nerve damage they cause. They also may look less serious to the uneducated eye because they do not ooze, and when they are pale they look more like normal skin than does the angry flush of superficial burns. Depending on their size and location, they may heal on their own, though not without leaving scars, or they may need to be closed with a skin graft, a thin sheet of healthy skin taken from an unburned area of the body.

Third-degree, or full-thickness, burns penetrate the entire dermis, down to the layer of fat that lies beneath the skin. These burns have a tough surface consisting of dead tissue called eschar (ess-kar), from the Greek *eschara* for "scab." The eschar may be soft, like a leather garment, or as hard as shoe leather. The color ranges from red and white to brown and black, again depending on the source of the heat and how long it was in contact with the skin surface. Scalds and steam will produce a red or white eschar whereas flame will toast it brown or char it black. As horrible as these burns look and as difficult as it is to believe, they are less painful than second-degree burns. That is because the nerves in the skin have been destroyed, along with everything else from the surface inward. A deep ache in adjacent tissues is characteristic of full-thickness burns, in contrast to the intense, burning pain of more superficial ones. The total destruction of the skin means full-thickness burns cannot heal on their own. The eschar and all underlying dead tissue must be removed and the wound covered with grafted skin. Even with skillful grafting, scarring is inevitable.

Burns can go even deeper, extending through the fatty layer and the fascia, the papery tissue between fat and muscle, all the way to the muscle itself, and even down to the bone. These very deep burns are sometimes referred to as fourth-degree. Amputation is often the only option.

These classifications make it sound fairly clear-cut, but even the most experienced burn specialist may not be able to tell right away how deep a burn is. Burns and the area around them undergo changes during the first few days, often getting worse rather than better. Furthermore, burns are rarely of uniform depth. More likely there will be a combination of superficial and deep second-degree burns, or both second- and third-degree, often adjacent to each other. Much depends on the thickness of the skin, which varies in different areas of the body, and whether one part was shielded by clothing or in some other way while another was directly exposed to the heat source.

Think of a burn as three concentric circles extending in three dimensions. This is a simplified scheme, but it gives a good idea of the landscape of a burn wound. Damage is greatest in the center, where heat has coagulated the proteins contained in flesh and blood. This is truly a dead zone, the damage irreversible. Moving outward and downward, tissues are injured but not destroyed. Blood flow in this zone is reduced or stopped entirely, a condition called ischemia, which may eventually lead to cell death and thereby extend the area of irreversible damage. The outermost and deepest zone sees little or no cell destruction, but a lot of swelling and inflammation, the body's way of screaming out for help when it is injured.

Burn degrees get a lot of attention, especially among the general public, but depth alone does not determine how severe a burn is, nor whether a burn patient will recover, with or without disfigurement and impairment. How much of the body surface is burned is equally important; in the early hours after a burn, it may be more so. It is central to all initial life-saving measures, as well as to subsequent treatment and prospects for survival and recovery. The larger the burn, the more rapidly and intensely the entire body responds to the assault, and the greater the need to intervene to reverse that response. For burns that require grafting, how much *unburned* skin is available will have a major impact on treatment decisions. Except in the case of first-degree burns, which do not really count, the focus tends to be on burn percentage more than degree. That is why you will hear burn specialists say, "She had a 40 percent burn," more frequently than, "She had second-degree burns."

In an adult, anything larger than 10 percent TBSA (total body surface area) is considered a major burn. In children and the elderly, a burn of any size is viewed as severe until proven otherwise. The quick and dirty way to estimate burn percentage is to use the rule of nines, which divides the body into parts, each representing 9 percent (or a multiple) of TBSA. For example, the arm of an adult is 9 percent, as is the head; the leg and back are each 18 percent. Smaller, scattered burns are measured by using the size of the patient's palm as equal to 0.5 percent of TBSA. The rule changes for children, whose heads are larger in proportion to their bodies and legs shorter than their trunks.

The Lund-Browder method, part of the legacy of Cocoanut Grove, aims for greater precision. At Boston City Hospital, Drs. Charles Lund and Newton Browder devised a way to map burns that divides the body into many more parts and assigns a percentage to each, ranging from 1 percent for the front of the neck and the genitals to 13 percent for the back down to the waist. Lund and Browder developed two separate diagrams, one for adults and one for children, reflecting the marked difference in body proportions. Their body surface area diagrams, still used today, show the front and back of the body in silhouette and marked off in segments, the percentage indicated for each. When the location of all burns is filled in and the areas added together, the result is the percentage TBSA.

Although a burn begins as an injury to the surface of the body, in many ways that is the least of the problem. We now know that when a burn is deeper than first degree and larger than 10 percent TBSA, the entire body responds to the injury; when it is larger than 20–30 percent, that response may be violent, complicated, and far-reaching. The body is caught up in a relentless cycle of assault and defense, of breakdown and repair, of destruction and regeneration. In its effort to restore balance, the human organism marshals all its resources, but at any point the balance could tip toward death.

* * *

There is a reason why people who treat burns refer to their patients as *sick* rather than *injured*. Dr. J. Long, a nineteenth-century pathologist, observed that "burns are not a simple problem, but a very complicated disease," and indeed a major burn is a progressive systemic disorder. What begins as an assault on the skin sets off a widespread riot throughout the body.

In the immediate aftermath of a severe burn, the gravest dangers are from shock and suffocation. These conditions are two sides of the same coin. The common denominator is oxygen—the lack of it. Neither the circulatory nor the respiratory system is doing an adequate job of delivering oxygen where it is needed.

Many people use the term "shock" loosely to describe an emotional state, a stunned response to stress or to a sudden and disturbing experience. In fact, shock is a physiological state that occurs when blood pressure is too low to carry oxygenated blood to all parts of the body, most commonly because the heart is not pumping out enough blood (cardiogenic shock) or because there is not enough blood for the heart to pump (hypovolemic shock). Whichever the case, the result is the same: not enough blood to meet the body's oxygen needs, which leads to cell death and eventually tissue damage and organ failure. Rapid but shallow breathing means more oxygen is needed; a rapid but weak pulse means the heart is laboring in vain to deliver the blood. The characteristic pale and clammy skin and drop in body temperature also hint that blood is not flowing as it should be. Often the toes and fingertips are bluish (cyanotic) and cold, signs that oxygenated blood is not reaching the extremities. Typically urine slows to a trickle or stops all together. Restlessness, extreme thirst, a blank expression, and dull, staring eyes with dilated pupils are other features of shock. A person in shock may be conscious or unconscious, and if conscious, may be in an altered mental state—highly agitated or apathetic, out of it, unresponsive.

Any number of things can bring about shock, among them traumatic injury, severe allergic reaction, drug overdose, dehydration, and toxins released by some kinds of bacteria. Shock may be more or less profound. It may be transitory or it may require aggressive intervention. It is sometimes, but not always, life threatening.

Hypovolemia is the most common type of shock, and loss of blood, through either external or internal bleeding, is the most common cause of hypovolemic shock. It is possible, however, to lose blood volume without bleeding at all. That is what happens with a burn. Burns involving as little as 15 percent TSBA put a person at risk for shock. Add inhalation injury to the mix, and the risk increases enormously, as do the potential complications. Burn shock develops when fluid leaks out of the blood vessels and into the surrounding tissues. This is not due to bleeding; there is no visible rupture in the blood vessels and it is not whole red blood that is lost. Although burns might be accompanied by broken bones and other traumatic injuries that do involve bleeding, something else is going on here.

After a burn, the vascular system turns into a giant sieve through which gushes plasma with its payload of serum (mostly water), proteins (including albumin, antibodies, complement molecules, clotting factors, and hormones), electrolytes (sodium, potassium, calcium, and magnesium, important regulators of heart, kidney, nerve, and muscle function), fats, sugars, and vitamins. The disruption takes place in the microvasculature—the capillaries—where gases and nutrients are exchanged between blood and the tissues it feeds. By design, the walls of these hair-thin vessels are semipermeable to permit the exchange, but in the minutes after a burn something occurs to make them highly permeable and the exchange becomes a one-way street. No longer is it only gases and nutrients that pass through the capillary walls; now colloids such as proteins and other particles suspended in blood are pushed out of the vasculature along with a torrent of fluid. At its most extreme, this hyper-permeability means that red blood cells are the only particles too large to pass through the capillary walls. Before long, most of the three to four quarts of plasma normally circulating throughout the body has spilled out into the intracellular space, where it does no good and can do only harm. With the loss of so much fluid, a condition called hemoconcentration develops as circulating blood becomes thick, slow moving, and prone to clotting.

The drop in blood volume causes a sharp drop in blood pressure and consequently a marked decrease in the amount of oxygenated blood pumped by the heart. The reduction in circulating plasma also robs the entire body of a crucial temperature regulator; combined with the loss of the skin's evaporative heat shield, this sends the body temperature plummeting.

And then there is edema, the ugly offspring of the fluid shift. When plasma leaks out of the vessels, it is not just sloshing around inside the body cavity. Some of it seeps from the wound, but most of it is taken up by the tissues like a sponge sopping up a spill. And like a sponge, the tissue expands. The waterlogged tissue further interferes with the movement of blood by pressing on blood vessels, narrowing and even collapsing them. At first, the swelling concentrates around the burned area. It begins almost immediately; within an hour, the water content of burned tissue may double. This is when blisters form, as plasma collects between dead and living skin layers in a superficial partial-thickness burn. It is also when victims of smoke inhalation suffocate, their throats swollen shut. After about one hour and continuing for the next twelve to twenty-four, the capillary leakage spreads, and with it the edema. The tissues with the richest blood supply and therefore the most capillaries—the lips, the eyelids, the mouth, the throat, the genitals—swell the quickest and the most. A burned hand looks like a catcher's mitt, a burned head like a beach ball. Within twenty-four hours, patients gain enormous amounts of weight, all of it from water. Burn victims have been known to puff up like the Michelin man, their bodies grotesquely swollen, their faces unrecognizable, their airways closed by engorged tissue.

The edema reaches its peak in about twenty-four hours, then gradually resolves. It takes a lot longer for the swelling to go down than it took to build up, so it may be a week or more before the watery excess seeps out of the wounds or is reabsorbed by the circulatory and lymphatic systems and carried to the kidneys for elimination as urine.

* * *

The physics of the fluid shift mostly has to do with hydrostatic and osmotic pressure, the push-pull that governs the behavior of liquids. When plasma is pushed out of the capillaries by hydrostatic pressure, dissolved sodium and other electrolytes go along with it. Plasma also contains albumin, a large-molecule protein that counteracts hydrostatic pressure, using osmosis to keep fluid inside the blood vessels or to pull it back in if it has escaped. In contrast to the electrolytes, albumin molecules cannot pass through the normal capillary membrane because they are too big. When the membrane is hyperpermeable, however, albumin can and does get through, pushed out along with the fluid and electrolytes. Once albumin is on the other side, there is no osmotic force to pull plasma back where it belongs.

The fluid shift begins at the site of the burn, but in all but the smallest burns it spreads to unburned and even distant parts of the body. Osmotic and hydrostatic pressure may drive the process, but the explanation for the complete disruption of the normal capillary barrier remains in the realm of the educated guess. The best guess is that something causes the cells that make up the capillary wall to contract, enlarging the space between them. That something probably has to do with a handful of specialized cells that are released as part of the inflammatory response.

The inflammatory response is a highly complex immune system mechanism, with a cast of thousands featuring prostaglandins and neuropeptides, polymorphonuclear leukocytes and macrophages, interleukins and interferons, a parade of growth factors, epinephrine and other stress hormones, and series of multifaceted chain reactions called cascades that bring specialized cells to the site of injury and activate them to release chemicals, which in turn signal others to turn on or turn off or send messages to bring more actors to the scene. There are agonists and antagonists, pro-inflammatory and anti-inflammatory agents, inhibitors and stimulators, each having a job to do but often working at cross purposes with others in the neighborhood. Some cause blood vessels to constrict; others make them dilate. Some clot, others dissolve. Some gobble up dead cells and spit out toxins, which poison healthy tissue in the area or wreak havoc downstream in the kidneys. Like most battlefields, a burn wound is a scene of chaos with more than its share of casualties from friendly fire.

A case in point is the disruption of the capillary barrier. No useful purpose is served when plasma spills out of the circulation. It is simply an unfortunate side effect of the activity of bradykinin, histamine, and leukotrienes, three chemicals brought to the site of injury for legitimate reasons. A stunning example of collateral damage is responsible for the fact that most burns get worse before they get better. Heat-injured tissue is subject to secondary assault from inflammatory mediators that happen to be toxic to tissues. The two biggest culprits are superoxide, a free radical that fractures cell DNA, and proteases, a class of enzymes that break down protein, which is, after all, what flesh is made of. A specific type of protease called metalloprotease is present and highly active in burn wounds. Its purpose is to digest damaged tissue, but an unintended consequence is that it also destroys growth factors and new tissue, thereby interfering with healing. Another consequence of the clash of armies is wound conversion, whereby a partial-thickness burn transforms itself into a full-thickness one without any additional exposure to fire or heat. This gruesome transformation takes place as the dead zone expands outward and downward. Burns are by their nature oxygen deprived (hypoxic) since blood flow is nonexistent in the central zone and greatly decreased in the ischemic zone. The progressive cell death that causes a wound to convert is a combination of hypoxia and another string of unintended consequences. Superoxide and metalloprotease do their damage, then macrophages and other scavenger cells arrive to gobble up destroyed tissue. Once these zealots get started, they do not know when to stop. And the wound bores deeper into the flesh.

The dominoes line up for miles, and they continue to topple long after the initial inflammatory response has subsided. Overactivation soon depletes the population of polymorphonuclear leukocytes and macrophages, leaving burn patients immunodeficient and highly vulnerable to infection. Something similar happens to clotting factors. Overabundant in the immediate postburn period, they are pretty much spent when they are needed to halt bleeding. The heart may suffer the effects of overwork or lack of oxygen, the lungs may fill with fluid or collapse. Prolonged hypoxia results in brain damage and may end in death. The gastrointestinal tract is prone to a constellation of disastrous events: the development of stress ulcers; the escape of bacteria that do important work within the confines of the digestive system but wreak havoc elsewhere in the body; paralysis of the gut that puts the entire GI tract out of commission; and abdominal compartment syndrome, in which pressure builds up in the abdominal cavity until, blimplike, it encroaches on the diaphragm and major blood vessels, impeding both breathing and circulation and resulting in injury to the heart, lungs, brain, and kidneys. Reduced blood flow is a problem throughout the body, with the areas most distant from the heart being last in line for delivery. Damage to the hands and feet, even when they have not been burned, is a real risk, which is why peripheral pulses are carefully monitored from the start.

The inflammatory response is self-limiting, and the hyperpermeable state seems to correct itself after a while, but that does not mean it can be allowed to run its course without intervention. Burn patients who do not get supportive care will suffer the consequences of shock, up to and including death. It is vitally important to replace fluid lost from the circulation and somehow keep it there, as well as siphon off the overflow in the rest of the body. The process is called fluid resuscitation; the key is a simple and relatively inexpensive cocktail of sodium, chloride, potassium, calcium, and lactate called lactated Ringer's solution. But there is a catch: As long as fluid continues to pour through the capillary walls, the fluid that is needed to combat shock ends up feeding edema, which compromises circulation, which contributes to shock, making for a particularly vicious cycle. The trick is to add enough fluid to restore circulation to near normal, but not so much that edema gets out of control.

* * *

The renowned French surgeon Guillaume Dupuytren was extraordinarily modern in his grasp of the pathophysiology of burns. His observations are like a flash of bright light before darkness descended once again on the subject. He is quoted as follows in the introduction to what was in the 1940s the standard work on burns: "The life of the patient may be endangered successively at four different periods, the period of irritation, period of inflammation, period of suppuration, and period of exhaustion." The author of the 1942 text goes on to observe, "In this first stage Dupuytren found internal congestion as the chief pathologic result. He explained this by the somewhat fantastic theory that 'the blood propelled internally by a general and sudden irritation, has made an effort under the influence of the excessive stimulus of the heart and of the vascular apparatus, to escape through all the free pores of the inner surfaces.'" The only thing fantastic about this is that Dupuytren could have been so close to the truth whereas a "modern" book published a hundred years later found him quaintly off the mark. We now know fluid from the blood *is* "propelled internally" (pushed from within the capillaries) and it *does* escape through "pores of the inner surfaces" (the gaps between the cells in the capillary walls). Perhaps when we gain a deeper understanding of the biochemical mechanism underlying the hyperpermeable transformation we will be able to stop the process sooner or prevent it from occurring in the first place. That in turn might minimize the fluid shift and banish the twin specters of burn shock and edema.

To that end, researchers are examining all the players in the inflammatory response in the hope of finding ways to prevent their release, inhibit their activity, or disrupt the cascade of events in which they have a role. Investigators have tried using histamine-receptor antagonists and inhibitors—drugs related to the antihistamines and mast cell stabilizers used for allergies—to block or minimize histamine's effect on the capillary wall. They have tried to dampen the inflammatory response with drugs related to those used for arthritis (anti-inflammatories). Because serotonin (more commonly thought of in connection with brain activity) also amplifies the activity of histamine and other inflammatory mediators, researchers are looking into the use of serotonin antagonists during the burn shock period. There is a possibility that protease inhibitors, not unlike those used to block viral replication in the treatment of HIV/AIDS, might short-circuit one of the complex chain reactions that fuel the inflammatory response, though the necessity of pretreating experimental animals would be a major obstacle in clinical practice, given the unplanned nature of a burn. Thus far, none of this research has borne fruit, and fluid resuscitation remains the best available strategy for riot control. But current theories about the fluid shift may someday be regarded as "fantastic" and pumping in fluid and siphoning off urine may seem as crude as the boiled frog and oak leaf tisanes of yesteryear.

One of the reasons why the Cocoanut Grove patients got so much less fluid than similarly injured patients get today is that estimates of what was lost were based on what could be wrung out of their bedsheets plus the assumed evaporative loss from the skin. Oliver Cope and Francis Moore's determination of how much fluid was lost from the circulation included fluid sequestered in the tissues, what they termed the "third space." Their formula was widely accepted, but disagreement persisted about what kind of replacement fluid was best.

The use of the term "formula" may be confusing. It refers not to the specific fluid or fluid combinations, but to the numerical formula used to derive the quantity, rate, and duration of fluid replacement. In the years following Cocoanut Grove, the principles and materials of fluid resuscitation were refined and revised. Body weight was added to surface area in determining the quantity of replacement fluid. For the most part, lactated Ringer's solution came to be used instead of normal saline because it more closely matches the electrolyte content of plasma and the lactate it contains helps normalize blood pH. But differences of opinion continued about both formulas and fluids, and especially about whether there was any advantage in adding albumin or some other colloidal substance to slow or halt the fluid shift.

Different fluids and combinations of fluids were proposed, as were variations on the rate and duration of infusion. Finally, in 1968, Charles Baxter and Tom Shires came up with what is known as the Parkland formula, named for Parkland Memorial Hospital in Dallas, Texas, where the two men worked. It called for infusion of 4 milliliters of lactated Ringer's times the patient's weight in kilograms times the percentage TBSA over the course of twenty-four hours, half the total to be given in the first eight hours, one-quarter in the next, and the final quarter in the third eight-hour period. Three years later, Jack Moncrief and Basil Pruitt, at the U.S. Army Institute of Surgical Research, proposed a modification of the so-called Brooke formula (developed in 1953 and named for the

Brooke Army Medical Center in San Antonio, Texas, where USAISR is located), which prescribes half as much fluid as the Parkland formula. Although the original Brooke formula used three parts Ringer's to one part colloid, neither Parkland nor the modified Brooke uses plasma, albumin, or any other colloid. The thinking was that these additives are at best wasted, and probably do more harm than good in the first twenty-four hours. As long as the cellular gates are open, the argument goes, the large molecules pour out of the circulation along with everything else; once the gates are closed, the molecules are on the outside, resisting the pull back through the capillary wall and thus slowing reabsorption of extravascular fluid.

Parkland is the formula most widely used today for adult burn patients, with the modified Brooke a close second, but they are far from the only ones. Some formulas combine lactated Ringer's with some or a lot of extra salt; some use plasma along with Ringer's, others use albumin or dextran, a polymerized polysaccharide molecule that can be made in different sizes. After detailing the pros and cons of each, the most recent American Burn Association Practice Guidelines concluded: "It is clear that all solutions reviewed are effective. . . . However, it makes no more sense to use a single fluid for all patients than it does to use one antibiotic for all infections. . . . The volume of fluid necessary to resuscitate burn patients depends on severity of injury, age, physiological status, and associated injury [especially inhalation injury]. Consequently, the volume predicted by a resuscitation formula commonly must be modified according to the individual's response to therapy." In other words, regardless of what you use and how you use it, pay attention to the clinical picture—how each individual patient is doing—and make adjustments as needed.

* * *

Resuscitation is a job for experts, not because it is an exact science but because it is not. If it were just a matter of hooking up an IV bag and dispensing a prescribed fluid according to a rigid formula, practically anyone could do it. But any formula is just a starting point and then, as Bob Droste puts it, "you adjust for reality." A plainspoken man with a sardonic edge, Bob has been resuscitating burn patients at Mass General since before there was a burn unit, and there is not much he does not know about how to do it.

Bob Droste is no one's idea of what a nurse looks like. His three-day beard is more salt than pepper and he covers what hair he has with one of those shower cap–style blue paper hats set at a jaunty angle and held in place with a Red Sox visor. He dresses for action, not appearance. His sense of humor runs to the ironic and his speech to the colorful. He obviously uses both to battle burnout, and for more than thirty years this strategy seems to have worked. He came to MGH in the early 1970s after serving in Vietnam as an army clinical specialist, the equivalent of a licensed practical nurse (LPN). "I thought I'd work in the Emergency Room, because that's most like what I'd been doing, but they said there was no opening." He was assigned to burns, with the possibility of transferring down to the Emergency Department after six months. "But then I sort of forgot why I wanted to do that, so here I am: saving lives and stamping out disease."

He had passed his twenty-fifth anniversary at Mass General when the hospital eliminated the LPN position. "I became an RN when they said they would throw me out." At an age when another kind of person would have hung up his hat, Bob Droste went back to school and got the degree he needed to keep doing the work he does so well.

"Ever since the first six months I worked here, I've taken care of the sickest patients. That's what I like to do. It works for me, and I'm pretty productive at it."

The sickest patients are the ones for whom resuscitation is a matter of life and death. As Bob says, "They die if you don't resuscitate them, so you have to make them be alive, otherwise there's no point" in all the rest of what Bob and his colleagues do to help burn patients heal.

Fluid has to be started as soon as possible, ideally before the burn victim even arrives at the hospital. Time is of the essence, and it could take a lot of time to get to a specialized burn center like the one at Mass General.

"The really important thing," Bob explains, "is the ABCs—the basic paramedic stuff—and then to get fluid running and get them here. The paramedics start dumping in fluid in the field. When they get to the ER, there's always somebody who knows what's going on with a big burn. First they look at the patient and make sure he's alive. And if the patient's alive, then they look at the burn."

By the time patients are turned over to the burn service, they will have had liters of fluid pumped into their system. "It varies from a very good resuscitation to crap, and we're used to that. We get a report on how much fluid they've received—in the ambulance, en route, in the ER—and we don't believe it. We expect it to be wrong, but it is a reasonable place to start. So we add what they tell us to the Parkland formula or whatever it is that is in vogue at the time, and we take it from there."

Nurses do not decide which formula to use; the doctor in charge does that. Most of the time, doctors write their orders according to standard protocols, but they may modify them to suit the specific circumstances or their own preference. At Mass General, the protocol is to begin with Parkland. But Parkland is not a magic formula; it is a ballpark estimate, a place to start.

Bob pulls out a handheld computer and shows how he gets to the ballpark. There is a body blank on the screen, the basic Lund-Browder map in pixels. He uses a stylus to mark the burned areas, taps in an estimate of the patient's weight, hits the enter key, and out comes the amount of fluid and the infusion rate. "We don't weigh the patients. There's no way you have time to do that. You can make a guess that's close enough for this. This is a pretty cool thing. I mean, it's not totally accurate, but it's close enough. Everything is an estimate, even TBSA. But if you have any experience and you fill it out carefully, you can end up with a pretty close percentage of TBSA. If you were in a ding-a-ling little hospital that didn't know about burns, you could do this and start your fluid resuscitation, and then send the patient on to us."

But, he adds, "It never works." For one thing, none of the formulas takes inhalation injury into account. The calculations are based on body surface area, and the surface of the lungs and respiratory tree is not factored in. Age and existing health problems make a difference too, Bob explains. Fluid loss in children tends to be greater than in adults, and because their normal blood volume is so much less than an adult's, each milliliter represents a larger proportion of the whole. Shock can develop very quickly in children, and can quickly lead to death. Edema also develops quickly and is often extreme. For these and many other reasons, Bob says, "Very young kids can be very unstable no matter what you do." Older people often have cardiovascular, respiratory, and kidney ailments that complicate fluid replacement and increase the risk of shock and the dangers of edema. "Old people do very poorly if they aren't resuscitated well and *very carefully* because too much fluid will hurt them as much as too little."

So it takes more than number crunching; it takes constant monitoring of a laundry list of indicators of how well the heart is pumping blood, where the blood is going, and how the entire body is withstanding the assault. Nurses follow the protocol, which includes a set of algorithms aimed at specific goals: If this happens, do that; if this does not happen, do that; if this value is too low, add that; if this one is to high, change that. "We set up the Parkland formula and we start to use it, but as soon as it isn't working, we switch," Bob says.

Urine and blood pressure are two main indicators of adequate fluid replacement. "They get up here and if they're not making any urine, we give them more fluid. If their pressure isn't good and you're giving a huge amount of fluid, sometimes you throw in some colloid." They watch to make sure the lungs are not filling with fluid. They monitor respiratory rate, body temperature, and cardiac output, and regularly check peripheral pulses and other signs that blood is getting to the extremities. They measure hematocrit to keep track of the red cell concentration of the blood. They need to stay on top of electrolyte balance and oxygen saturation of the blood, and make sure blood pH does not dip into the acidic range, a sign of excess carbon dioxide. And they do it all nonstop, hour after hour, until the patient is stable. Even after that, a patient may get replacement fluid for weeks. "We don't call it resuscitation," Bob explains, "but they still need fluid."

And then, just when it seems the crisis is over, another riot breaks out.

* * *

The body is an internal combustion engine. It uses oxygen to burn fuel from food, which produces heat and, as waste product, carbon dioxide. A burn wound is an engine run amok. Driven by the inflammatory response, it consumes prodigious amounts of fuel and oxygen, and produces carbon dioxide beyond the body's capacity to dispose of it, in the course of which it bumps the pH of blood into the acid range. It also generates a lot of heat, which leads to fever. Hypermetabolism, as this phenomenon is called, begins a day or two after the shock period ends, steadily increasing over the next week and lasting at least until the wound has healed. In some cases, accelerated metabolic rate and increased energy demand continue for as long as nine months to a year.

Like the mechanisms of the fluid shift, the driving force behind hypermetabolism is highly complex, the causes not fully understood, and the strategies for minimizing its effects imperfect. The course and the consequences are nothing short of bizarre. Some call the first twenty-four hours after a burn the "honeymoon phase." But this wild "honeymoon" is nothing compared to the rocky road ahead—Dupuytren's second period, which he said begins around the third day after the burn. The French surgeon was right: Inflammation is at the bottom of it. And like the inflammatory response that spawns it, hypermetabolism is an overreaction that quickly spins out of control. In the first two days after a burn, metabolism slows down. During this so-called ebb phase, temperature, heart rate, and energy expenditure are lower than normal. Then suddenly the body catapults from one extreme to the other. Instead of low cardiac output, the heart starts pumping out blood at about double its usual volume. Anemia takes the place of hemoconcentration, hypothermia yields to fever, and the huge water weight gain of edema gives way to muscle-wasting weight loss that can exceed two pounds a day.

The last is probably the weirdest thing about this chaotic state. Something goes seriously wrong with the way the body translates the language of hormones, and instead of burning fat as its primary source of energy, it begins to burn protein. The results are catastrophic.

Metabolism encompasses all the chemical processes in the body that require energy. Catabolism, or destructive metabolism, is the breakdown of tissues. Anabolism, or constructive metabolism, is the process of synthesizing protein. Protein is the reservoir of amino acids, which are used to construct, maintain, and repair cells and tissues, and to manufacture hormones, enzymes, and immune system components. Most of the body's protein is in the skeletal muscles. It is the meat we are made of. The body's principal source of energy is fat, most of which is stored in pockets beneath the skin. Fat is an extremely efficient fuel source. It is easy to process from food for storage, easy to convert to fuel, and gram for gram it provides more than twice as much energy as protein (9 calories versus 4 calories). Protein is not a particularly efficient source of energy, nor is it meant to be used as such. It takes a lot of energy just to break it down into usable form, and the effort is not rewarded with much in the way of fuel. Normally fat supplies 95 percent of the body's energy needs and protein only 5 percent. In the throes of hypermetabolism, however, not only is more fuel burned, but about 20 percent of what is burned comes from protein. This might not seem like a big deal until you consider that the average adult who is neither muscle-bound nor obese carries about 40,000 calories of energy stored as protein compared with 150,000 calories of stored fat. It does not take long to deplete the protein depot—the lean body mass—and it does not take much depletion to make a big difference.

Losing 10 percent of lean body mass impairs immunity and increases the risk of infection; a 20 percent loss slows wound healing—bad in a burn patient. By the time 30 percent is lost, healing grinds to a halt, sepsis and pneumonia take hold, and muscle wasting is so pronounced that the patient is too weak to even sit up. A 40 percent loss of lean mass is fatal.

The vortex of the hypermetabolic storm is the burn wound. As well as anyone understands, the trouble begins when stress hormones are released at the height of the inflammatory response. These are the "fight or flight" chemicals, the ones that call for a rise in the metabolic rate to generate the energy to challenge the saber-tooth tiger or to cut and run. But this time there is no tiger nor any other external threat. Instead, there is a wound that needs oxygen and amino acids if it is to heal, and it has a hard time getting either. A revved-up metabolism hogs the raw materials needed for healing and dumps its waste products into the circulation, adding to the demands on the already overburdened heart, lungs, and kidneys.

Ordinarily, the metabolic rate subsides once an emergency is over. If it looks like there will not be enough readily available fuel to maintain metabolism at its usual rate, it will even gear down below normal. Reducing demand in the face of reduced supply is the body's way of adapting to famine and the reason why it is so hard to lose more than a few pounds simply by eating less. But burn hypermetabolism overrides the message to slow down, and it just keeps going and going.

It does not take a huge burn to throw the metabolism into overdrive. A burn as small as 10 percent TBSA can increase the metabolic rate (resting energy expenditure, or REE, expressed in kilocalories or, in everyday usage, calories) by 15 percent. Burns larger than 50 percent can double REE, consuming 2,000–3,000 calories more than normal, every single day. If those calories are not provided by food, the body will cannibalize itself to get them.

The need to supplement caloric intake in burn patients, and the significant challenges in doing so, were addressed by doctors treating the Cocoanut Grove victims, at both Boston City and Mass General hospitals. Since that time, much has been learned about how much and what kind of nutritional support burn patients need. Death by starvation is no longer a common complication of severe burns, but hypermetabolism is still a force to reckon with.

As with the fluid shift, investigators are trying to better understand the underlying causes of hypermetabolism in an effort to prevent it rather than simply counteract it. They hope to develop pharmaceutical and other means to alter the biochemical activity that triggers and perpetuates metabolic acceleration. Some are studying the effects of various chemicals—growth factors, beta-blockers, and such modulators of inflammation and hormone activity as steroids and insulin. Others are focusing on the cellular and even molecular levels in search of what makes burns so energy demanding. At the present time, feeding the monster is the best way to fight hypermetabolism. Burn patients who can take food by mouth are encouraged to eat like there is no tomorrow. Those who cannot are given calorie-dense formulas through a stomach tube or, if that is not possible, intravenously. Because the body's need to fight low core temperature contributes to hypermetabolism, keeping patients warm so they do not have to expend energy to warm themselves helps apply the brakes to the metabolic engine. Supporting their respiration ensures they get enough oxygen and helps the lungs expel carbon dioxide. Making sure they get enough fluid helps dilute the kidney-toxic waste products of catabolism. And because this metabolic derangement originates with the burn wound itself, closing it as quickly as possible will begin to shift the engine into reverse.

John F. Burke, MD, the founding director of the burn treatment center at Mass General, observed, "It is important to emphasize that the hallmark of a burn injury is the length of the illness. A burn is not an injury that is over in a day or two. The acute illness covers not days but weeks or even months. One cannot think about the therapeutic activities that go on in the first day or two of the burn injury in isolation. One has to think about them in their entire context of the long pull, so to speak."

That long pull is an enormously complicated undertaking. It requires constant monitoring, balancing conflicting requirements, and the efforts of experts in a multitude of disciplines who understand the fiendish behavior of a major burn. That is why burn patients should be cared for in a specialized burn center, where everyone on the team knows what to expect and how to handle the unexpected. There are 139 specialized burn centers in the United States, and one of the best of them is at Massachusetts General Hospital. People burned as badly as Dan O'Shea can do no better than to end up there. The sooner they get there, the better.

CHAPTER THREE

The General

The ambulance made a right on North Grove Street, screamed across the empty Parkman Street intersection, and slowed only slightly as it took the curve of the circular drive at the entrance to Massachusetts General Hospital. It was 4:52 Saturday morning, January 31, 1998.

The right side of the drive is for taxis and private cars; the far left is reserved for ambulances. At any given time there may be three or as many as a dozen of them parked perpendicular to the curb. The ambulance nosed into the vacant slot nearest the canopy marked *Emergency*. As the back doors flew open, the paramedics pulled out the collapsible gurney with its blanket-wrapped cargo and aimed it toward the wide doorway. One steered while the other ran alongside, rhythmically "bagging" their patient as the wheels rattled over the terra cotta tiles that line the corridor of the Vincent building.

They were expecting a fire victim in the Emergency Department even though it had been less than ten minutes since the ambulance pulled up outside the burning building and just five minutes since it left the scene. As soon as it was en route, the ambulance alerted emergency admitting at MGH through a radio communications link that can connect any EMS vehicle with any hospital in the city. When a beige phone rings in the triage office, it closes a loop that begins with a call to 911.

Whenever there is a fire, building collapse, vehicle pileup, boating accident, shooting, stabbing, suicide attempt, or any of a dozen other ways to get badly hurt in greater Boston, word goes out from the "Turret," the hub of the metro area Central Medical Emergency Dispatch (CMED) system. Occupying a full floor of Boston police headquarters on Tremont Street, the Turret houses the communications centers for both the police and Emergency Medical Services, the agency responsible for ambulance and rescue operations in Region 4—the city of Boston, including Logan International Airport, and sixty-two surrounding towns and cities. Central communications for the fire department is located in the South End, but all three are connected electronically. The instant an operator logs a 911 call into the computer, the CMED network relays the details to all the city services that need to be involved. Every EMS-licensed ambulance in Region 4 is equipped with a global positioning system, making its whereabouts known to CMED and enabling rapid deployment of the closest available ambulance to the scene and then to the closest hospital equipped to deal with the situation. Provided, of course, that hospital is not on divert.

On divert. Those are dreaded words in a city the size of Boston, where hospital emergency rooms are often filled to overflowing. Federal law requires hospitals to treat anyone who arrives in need of emergency care. That includes walk-ins, who cannot be barred no matter how swamped the emergency room might be, though they may wait hours for treatment. The purpose of diverting ambulances is to make sure they do not deliver patients to hospitals that have reached capacity. According to Boston EMS rules, only two city hospitals can be on divert at one time, cold comfort when the nearest open ER is across town and the ambulance has to make its way through the nightmare that is Boston traffic. CMED ensures that ambulance drivers know which hospitals are on divert, but even with regular bulletins, a diversion can tip the balance from life to death.

* * *

If you get burned within a 150-mile radius of Boston, Massachusetts General Hospital is where you want to be taken. Known to Boston residents as "the General," it is the largest hospital in New England, and the city's biggest nongovernmental employer. It has small outpatient satellites scattered throughout the metropolitan region, but its main location covers about sixteen and a half acres on the south bank of the Charles River, bounded by Cambridge, Blossom, Charles, and Parkman streets. Four interconnected towers envelope a hodgepodge of smaller structures in the typical sprawl of institutions that have outgrown their original footprint over time. The centerpiece is the neoclassical Bulfinch building, which housed the entire hospital when it opened in 1811. Its stately Ionic columns preside over a grassy quadrangle that gives the complex the look of a college campus.

Indeed, it is a place of learning as well as healing. Some consider it the nation's premier academic medical center. It has a long and illustrious history, of which it is conspicuously proud. With more than $300 million devoted to research annually, the General boasts the largest hospital-based research program in the United States. It is the main teaching hospital of Harvard Medical School, and all members of the medical staff hold appointments on the Harvard faculty. Its world-renowned residency program turns out some of the best-qualified doctors in the country, if not the world. It is also a designated Level I regional trauma and burn center.

Trauma centers are classified as Levels I–IV. Strictly speaking, burns are trauma—bodily injury due to an external force—albeit an extremely complex subtype. In accordance with the U.S. Public Health Service Model Trauma Care System Plan, the state EMS authority designates facilities that meet established standards as trauma centers. The specific standards followed throughout the United States were developed under the auspices of the American College of Surgeons Committee on Trauma (ASCOT). The purpose of ranking the centers is to allocate costly resources in a manner that provides the most comprehensive care required according to patient needs without duplicating expensive services and equipment in some localities while leaving others underserved. This is particularly important for burns. Most burns that cannot be dealt with at home or in a doctor's office can be treated on an outpatient basis. Severe burns, however, require the specialized care and highly experienced personnel found at a burn center verified by the American Burn Association.

Level I trauma centers, of which the General is a prime example, are the hospitals where the most critically injured patients end up, through direct admission or transfer from smaller and less well-equipped facilities. They are usually large university-based hospitals that have the personnel and resources to deliver a 24/7 coordinated response to critical and intensive care needs. According to the ASCOT criteria, a Level I facility must be able to serve an entire region, with a staff of board-certified trauma surgeons and critical care nurses, operating rooms open and operational around the clock, state-of-the-art imaging and monitoring equipment, and rescue and response teams capable of providing prehospital care and fast transport.

Level II centers provide initial trauma care, which may be all a patient needs, but they often transfer patients to a Level I center if the injuries are complex and require comprehensive care beyond Level II capabilities and resources. Level III centers are usually community-based hospitals, staffed by at least one general surgeon, that provide prompt assessment, resuscitation, and emergency care; once patients are stabilized, they are transferred to a Level I or Level II facility, depending on the care they need. Level IV centers are located in remote and rural areas where there are no other trauma centers available. They may be hospitals or clinics, with or without physicians, but they are required to have personnel trained in advanced trauma life support (ATLS), the first aid techniques for life-threatening traumatic injuries. This hierarchical plan is central to regional and national emergency preparedness, be it for natural disasters or enemy attacks.

When it all works as planned, trauma and burn victims are rapidly evaluated wherever they enter the system; if the receiving facility is unable to provide the care required, they are transferred to the appropriate higher level. This requires excellent communication among trauma centers and a full understanding of patient needs at all medical facilities. When it comes to burns, that is where things can break down.

Emergency room and other medical personnel who lack experience and expertise in burns may not know how to evaluate the severity of the injury. They may mistake a large but superficial burn for something more serious or, more dangerously, may mistake a small but very deep wound, such as is typical of a high-voltage electrical burn, for a relatively minor matter. They may be unaware that even apparently small burns on the face, hands, or genital area require expert attention. They may not recognize the presence of inhalation injury or understand how perilous it is. They may wrap a patient in a wet sheet or bandages in a misguided attempt to cool the burned area, risking severe and potentially fatal hypothermia. They may be unaware of the importance of fluid resuscitation and lose a patient to burn shock. They may not realize that burns to the torso can prevent chest expansion and thus hamper already compromised breathing or that even small circumferential burns that wrap completely around an arm or a leg or a neck can cut off blood circulation. They may focus on the burn itself and fail to deal with the catastrophic systemic response. They may be unaware that a burn victim who is very young or very old is at far greater risk of dying from apparently minor burns. Any one of these and a hundred more gaps in knowledge could result in a "poor outcome" due to a hospital's failure to transfer a burn victim to a facility capable of saving lives that would otherwise be lost.

Burn and trauma patients arrive at the General in a multitude of ways. They may be direct admits like Dan O'Shea or may be transferred from hospitals elsewhere in the city, state, or New England region. They come by ambulance, by private car, by foot. Some even come by helicopter, touching down on the helipad atop the hospital's Blake building. And they come in droves. Close to 75,000 patients pass through the Emergency Department each year, and its facilities dwarf those of many a small hospital. Indeed, it would be the greatest of understatements to refer to this sprawling complex as an "ER."

The treatment area is a vast echoing space shaped like a right triangle. At the apex is the triage nurse's station and the rapid diagnostic unit, where waiting patients are assessed and triaged to ensure that everyone gets appropriate treatment and the most gravely ill or injured are treated first. Along the left side, the hypotenuse, are four open trauma bays, each with one bed and an array of equipment curtained off from its neighbor. The area widens at the back. To the left is a corridor with five closed bays, visible through sliding glass doors. One of the closed rooms can accommodate two patients and two trauma teams, bringing the count up to a ten-bed acute care unit fully equipped to handle shock trauma, cardiac, and stroke acute care, and burn resuscitation. To the right is the major multipurpose area, a large open space with twenty stations furnished with reclining examination chairs, for less seriously ill or injured patients who may nonetheless need to lie down to receive care. Beyond that is a dedicated emergency radiology facility, with state-of-the-art imaging technology. Children are taken care of in a separate four-bed pediatric emergency unit, and there is an emergency psych unit on the opposite side of the waiting room.

* * *

Things were relatively quiet when Dan O'Shea was wheeled into the Emergency Department. No one knew his name. Someone was working on the ID and locating the family, but for the moment he was an "unknown."

By 5:00 A.M., he was in one of the closed bays, dubbed the "warm room." The ambient temperature was approaching the high nineties with the help of overhead lights and portable "french fry" lamps arranged around the bed. The sliding glass doors were closed to retain heat and exclude the bacterial, fungal, and viral interlopers that populate every hospital, ready to attack those whose skin has been burned away.

All patients in the Emergency Department are the responsibility of the trauma service. Dan would not become a burn center patient until he got upstairs to the burn unit. The burn resident on call had been notified, and she in turn had beeped the on-call attending, but there was no time or need to wait. The trauma team knew what to do and went to work immediately.

The team moved expertly through an intricate choreography, each member executing well-practiced steps. Even before Dan was lifted onto the treatment bed, the basic drill began. The ABCs—airway, breathing, circulation—were repeated. Thanks to the ET tube inserted while he was in the ambulance, his airway was open and he could be rapidly connected to a respirator, which took over the job of breathing. The trauma resident checked him for spinal cord and head injury, puncture wounds, internal ruptures and bleeding, all of which are potentially life-threatening and would have to be treated immediately. The paramedics did a rapid primary survey when they pulled Dan from the fire, and their negative finding was borne out: No trauma to complicate the burn injury, though the burns were more than enough to kill him.

He wore his skin like a tattered shirt, a patchwork of red and white, black and brown covering his arms and torso. His hands looked like they were gloved in rags. His head was outsized and unnaturally round. His face was a lurid red mask, the skin shiny and taut, his hair and the tops of his ears singed, his eyelashes and eyebrows swept away by the flames. His lips were swollen into a wordless O around the ET tube. Miraculously, there were no burns on his lower body, which was protected by the heavy fabric of the blue jeans he was wearing when he passed out the night before. That healthy skin would be needed to replace what the fire had burned away.

The incident report handed over by the ambulance crew said Dan got a liter of normal saline while en route. The trauma resident subtracted that amount from the Parkland formula and hooked up the IV line already in place to a pump dispensing lactated Ringer's. Fortunately, Dan's pulses were strong all the way to his fingers and toes, evidence of peripheral perfusion, or blood flow to his extremities. Shock was not an immediate threat, and it could be kept at bay with continued fluid and respiratory support, careful monitoring, and more than a bit of luck.

The next most urgent among urgent tasks was to secure additional venous and arterial access. Throughout the life-saving period and for months afterward, fluids and medication would have to go in, and blood would have to come out of this ravaged body. Dan's vital signs—how well his lungs, heart, kidneys, and circulatory system were functioning—would have to be monitored closely. Blood gases and blood chemistry, systemic blood pressure, and central venous pressure would require easy access to his blood vessels. The best way to ensure that was by inserting flexible plastic catheters through his skin and into his veins and arteries. Not only would that save endless needle sticks, it would continually feed vital information to the beeping monitor standing watch over him. Getting arterial and venous lines in place is always a race against time. In Dan's case, edema had already begun to develop and would soon spread, obscuring his anatomic features, the joints and bony prominences lost in the tissue ballooning around them.

The first and second choices for central line placement are the neck and just below the collarbone, tapping into veins that flow into the superior vena cava, the huge vein that carries spent blood from the upper body into the right chamber of the heart. The resident rejected both sites on Dan because it would mean going in through burned tissue, always the last choice. He ruled out the arms for the arterial line because they too were burned. Instead he placed both venous and arterial lines in the femoral vessels, entering near the groin. A technician wheeled in a portable X-ray machine the size of a dishwasher and took a series of quick shots to ensure the lines were accurately placed. A nurse drew blood from the A-line and sent it off to the lab for a reading on blood gases, and even more important at the moment, carboxyhemoglobin level, an indicator of how much deadly carbon monoxide was circulating in Dan's blood.

A nurse swabbed a square of unburned buttock and shot in a tetanus booster to protect Dan from a deadly microbe that can sneak in through a burn wound and invade his body. She slipped a temperature probe into his rectum and checked the instant readout on the monitor. His temp was hovering around 94 degrees Fahrenheit, even though he had been covered with a sterile sheet and a blanket throughout his transport. Preventing it from falling any lower was crucial. Easing it up to normal was an effort that had to be made without getting in the way of everything else that needed to be done. The Ringer's solution had been heated to 102.5 degrees Fahrenheit, warming him from the inside, but heat was still venting through his shredded skin. The paramedics had cut off his clothes at the fire scene to keep them from choking his circulation as his flesh began to swell, and his burns would not be "clothed" in dressings until the burn attending had a chance to take a good look at them. In the meantime, everyone was sweating under the french fry lights except Dan, who was covered wherever possible with special air-heated pads called Bair Huggers.

No one will ever know how much Dan was aware of what was happening to him—the poking, the prodding, the pain. He drifted in and out of consciousness and seemed agitated at times, but that is not the same as being aware. Any movement can be attributed to reflex more than anything else. For all intents and purposes, he was completely out of it. An IV line was delivering morphine and Versed, the first for pain, the second to quell anxiety and induce amnesia. He would be on a continuous drip of these powerful drugs for a long time to come. Burn patients spend the acute period so heavily sedated that they are unlikely to remember anything of that horrible time. At least that is the goal, for the sake of the patients and those who love them.

* * *

If you can walk, you get to the Emergency Department by taking a right off the Vincent corridor just before the stretcher entrance. You will be facing a high semicircular desk, behind which sits a receptionist charged with answering phones, interacting with ambulatory patients and anxious family members, and paging emergency personnel over the PA system. At times these tasks are handled by a woman of indeterminate age with an arctic manner and long scarlet fingernails worthy of a Chinese empress. Peggy O'Shea was too distraught to notice fingernails or attitude if any was on display as she rushed to the desk and blurted out, "I'm Dan O'Shea's mother."

"We don't have a Dan O'Shea."

"But you must have. He's just been admitted . . . he was in a fire."

"Oh, that's the John Doe."

A wave of dread washed over Peggy. Daniel Patrick O'Shea, her thirty-one-year-old son, her youngest, was a nameless case, a body on a stretcher, a man no one knew. And she and Jack did not know if their son was even alive.

If it is possible to talk about luck in this situation, the O'Sheas' luck began to turn as soon as they met John Schulz, the burn attending on call that weekend. He is a man of skill and compassion, a straight shooter and a clear communicator. If you have to hear bad news, Schulz is the kind of guy you want to hear it from. And if it is hope you need to cling to, he will keep you company on that rough ride.

Schulz was the newest member of Burn Associates, the group of three attending surgeons—Rob Sheridan and Colleen Ryan are the other two—who codirect and care for patients on the adult burn service at the General, having come onboard less than a year earlier. He strode into the waiting area and introduced himself to Peggy and Jack. Sweaty in his green scrubs, his face mask hanging around his neck, he looked more like a med student than a savior, but Peggy and Jack quickly came to think of him that way.

A long tall Texan, he stands well over six feet in his cowboy boots. His elongated, narrow head is topped by a shiny dome, part of it shaved close to match the part denuded by nature. Heavy-rimmed glasses frame an intelligent face, but any trace of gravity is dispelled by the chaw of nicotine gum tucked in the corner of his mouth. He swears he is not and never has been a smoker, that the gum replaces a long habit of chewing tobacco. You tend to believe him, as you tend to believe everything he says.

Schulz is in his early forties, but is one of those people who remain boyish despite an exalted position and a bald pate. His down-home manner belies his obvious brilliance. He made his way to the General by way of Texas A&M. He picked up a PhD in biochemistry from Harvard and spent a year as a postdoctoral fellow in physiology at Tufts before getting his MD at Yale, where he was honored as "Outstanding Graduate Student in Surgery." He came to MGH as a surgical resident and remains as an attending surgeon on both the trauma and burn services. He also serves on the surgical staff at the Shriners Hospital for Children, around the corner on Blossom Street, and teaches at Harvard Medical School. With all that, he is refreshingly free of the swagger and aloofness often affected by surgeons.

Jack and Peggy noticed that immediately. Part of it was his deep voice, which thrums in counterpoint to the familiar flattened *a*'s and absent *r*'s of Beantown. He drops his *g*'s in an aw-shucks, country boy way and inflects his voice a lot, making the most of a rich palate of pauses, emphases, question marks. It is a friendly voice—he is an approachable person and the point of entry seems to be his voice.

Schulz led the way to a small room set aside for talks like the one he would be having with Jack and Peggy, when families need to be told their loved one is in grave danger and might not make it. He had had these conversations before and he will have them many times again before he lays down his scalpel for good, but they will never get any easier. What keeps Schulz going is the possibility that a patient like Dan O'Shea will pull through, that someone who looks like he is dying will someday walk into the outpatient clinic with a smile on his face and just a few telltale scars. But right then, he had a very sick patient on his hands and he had to do his best to explain to this shell-shocked couple what was happening back in the trauma bay, what it meant, and what he wanted their permission to do.

* * *

The first priority in the Emergency Department is to save a life. There is no legal requirement to get the consent of either patient or next of kin. Indeed, it may take time to identify the patient and locate next of kin, time no one can afford. A social worker and case manager work with the police and other agencies to connect the dots, but in the meantime, the hospital serves as the patient's proxy and undertakes all necessary measures to preserve life. From a legal standpoint, it is easier to initiate care than to withdraw it, even when care is futile.

With this one it was touch and go. Schulz could see that as soon as he examined Dan's scorched airway. The trauma team was deep in its work when he arrived at 5:10. He had been beeped as soon as Dan hit the trauma bay and somehow managed, as he always does, to be awake, alert, and ready for action within ten minutes of getting the call. He claims the secret is his "on-call" sleeping posture—flat on his back in a brightly lit room—and a cup of cold coffee he keeps on the kitchen counter. As he joined the beehive of activity in the warm room, he noted that everything that needed to be done was being done: The ambulance crew had done their job perfectly, the trauma team had everything in hand, the patient was on the vent, and fluid resuscitation was proceeding.

At 5:28, Schulz stepped in to perform a bronchoscopy, peering into Dan's lungs through a fiber-optic instrument threaded down his throat. What he saw was not good. The note he dictated later says it all: "Bronchoscopy showed vast quantities of carbonaceous material; entire airway is paved with soot." A sure sign of smoke inhalation. Not that it came as a surprise. Anyone found unconscious in a fire in an enclosed space is certain to have it, but inhalation injury just about doubles the chance of a person's dying from burns and it always complicates treatment and recovery, starting with fluid resuscitation and continuing for months and even years. Even without a burn, Dan's condition would be critical. If he lived long enough to have a plan of care, it would be driven by the state of his respiratory system.

The other shoe dropped when the blood report came back from the lab. Dan's carboxyhemoglobin level was 46.2 percent: disastrously high. Back in the tiny South Boston bedroom, the fire had quickly consumed most of the available oxygen and the resulting incomplete combustion generated carbon monoxide, one oxygen atom short of the carbon dioxide we normally exhale. As Dan lay slumped against his smoldering mattress, there was nothing left to inhale but carbon monoxide, a poisonous gas that has a binding affinity to hemoglobin, the oxygen carrier in red blood cells, nearly 250 times greater than oxygen. It cripples and kills by elbowing out the oxygen in the river of blood flowing throughout the body and to the brain. If Dan survived, he could end up severely brain damaged. Schulz consulted with colleagues, then put in a call to the

Hyperbaric Medicine Center next door at Massachusetts Eye and Ear Infirmary: An HBO chamber was needed, fast.

While they were waiting for the transfer to be arranged, Schulz turned his attention to the burns, systematically marking their location on the burn diagram. The version of Lund-Browder they use at the General shows the outline of an androgynous adult body, front and back, with insets focusing on critical areas—the male and female genitalia and the left and right underarms, or axillae, where contracted scar tissue can immobilize an entire arm. He examined every inch of Dan's body, from head to toe, front to back, and drew a crosshatch pattern wherever he saw a burn: the head, including both ears; face, including both eyes; neck, upper chest, both shoulders, arms, hands. He cross-hatched the right axilla in the inset. Fortunately, none of the burns was circumferential, the tourniquet-like wraparound configuration that requires immediate escharotomy, literally cutting through the eschar and beyond to permit expansion as everything beneath the rigid surface begins to swell. To Schulz's practiced eye, the burns looked partial thickness, but it would take a few days to be sure how deep they were. He called them second degree for now. Toting up the cross-hatched areas, he estimated the burns at 35 percent. Add that to the internal damage to the respiratory tract and he was looking at a very sick patient.

Schulz knew the man's parents were in the waiting area, and he slipped out to meet with them while the rest of the team continued to work. A neurologist was on the way to check the patient's brain function and an ophthalmologist needed to examine his eyes for possible corneal damage.

* * *

Sitting with the O'Sheas in the private family room, Schulz told them that their son was badly burned. If they were able to pull him through the next day or two he would need surgery, lots of skin grafts, and he would have scars, no question about it. But that was all in the uncertain future. The inhalation injury was immediately threatening and the carbon monoxide poisoning, well, no one really knew what the long-term effects might be. He explained that the team wanted to try hyperbaric treatment, high-pressure oxygen that could speed the clearance of carbon monoxide from his system. He asked for, and got, their consent.

The O'Sheas were listening carefully but had a hard time taking it all in. It felt like they had been sleepwalking through the four hours since they got the news. Years later, they might disagree about some of the details: "Did this happen the first day or was it the day after?" "Was it Dr. Schulz or one of the other doctors who told us that?" "He said it was like a ten-pack-a-day smoker." "I didn't hear him say that." But they do not differ in their first impression of John Schulz. "Dr. Schulz has a unique way of dealing with people that is very understanding, compassionate. He really knows how to settle you down." That is how Jack puts it. Peggy adds, "He's not emotional, but he's still concerned. And I think he tells it exactly how he sees it." What they remember most clearly is "Daniel was in very critical condition, he had smoke inhalation, and his carbon monoxide level was the highest that they'd ever seen, or something. And, you know, there was a chance he wouldn't survive this." But behind those words what Jack heard John Schulz say was, "I'm going to get this kid through if there's anything possible I can do."

Gently, Schulz steered the conversation to Dan's history, information needed in planning his treatment. Name, age, preexisting medical conditions, medications he was taking, previous surgeries and major illnesses. As they answered his questions, Jack and Peggy often finished each other's sentences, the broad Boston accent they share lending the quality of a fugue. Peggy began bravely, but when tears overtook her, she would cover her mouth with one hand while gesturing with the other. Jack would assume the narrative, gallantly shifting attention away from his wife. Finally, Peggy drew a deep breath and told Dr. Schulz that her son had a drinking problem. She knew he had to know this. But did he have to know about the other thing, what she and Jack call his "lifestyle"? She shot Jack a look and then decided that her son's homosexuality could not possibly matter right then.

While Schulz was out talking to his parents, Dan was being fitted with more tubes and wires to hook him up to the battery of pumps and monitors that would follow him wherever he went. A nasogastric tube was poked in one nostril and slid down into his stomach, first to vent any gas that could turn the stomach into a painful balloon and then to carry liquid nutrition to feed his hungry wounds and keep his bowels from shutting down. A Foley catheter was carefully slipped through his urethra and attached to a bag calibrated to measure his urine output. EKG sensors were clipped to his chest and a pulse oximeter capped one of his toes. Finally he was ready to be wrapped in gauze. And all the while, lactated Ringer's was being pumped into him. His blood pressure was holding steady, but his body was swelling, right on schedule.

Schulz told the O'Sheas that the nurses were dressing Dan's burns and preparing him for transfer to Mass Eye and Ear. It was a good time, he said, to take them back to see him. It might be frightening, but he promised to be right there with them. Jack took Peggy's hand as they followed the tall doctor to the warm room. It was 9:20 A.M.

Nothing Schulz told them could have prepared Jack and Peggy for the sight of their son. He was conscious, or seemed to be. His head was huge, his face bright red, and his eyes wild. He rose up from the table he was lying on and reached out his arms, beseeching them. The endotracheal tube made it impossible for him to speak, but the image is seared into their memory, and four years later they still insisted they heard him say: "Help me, help me."

Those haunting words were the last they would hear from him for nearly two months.

CHAPTER FOUR

The Very Young, the Very Old, the Drunk, and the Stupid

Dan O'Shea was drunk the night he got burned. That grants him the dubious distinction of membership in one of the largest classes of burn victims. It is easy for a person whose mind is clouded by alcohol or drugs to accidentally start a fire; it is difficult for a person whose judgment and mobility are impaired to escape. In addition to substance abuse, mental illness and out-and-out stupidity account for a disproportionate number of burns and a large proportion of those that require hospitalization.

The stupidity seems boundless. A do-it-yourselfer takes a cigarette break while applying polyurethane to the bedroom floor. A kid doing chores strikes a match to check the gas level in the lawn mower. An amateur car mechanic fiddles with some wires behind the dashboard and the airbag discharges sodium azide propellant, inflicting a ghastly chemical burn to the eyes and face. Candles and lacy curtains, fireplaces and shag rugs, curling irons and polyester bathrobes, paint fumes and pilot lights, hair spray and cigarettes, flimsy foil roasting pans weighted with turkey and sizzling gravy, charcoal lighter, floor wax, petroleum-based cleaning products, kerosene heaters, furniture varnish. Take a look around the house, under the kitchen sink, above the stove, in the basement workshop. Read a few labels. Think about it. The average home and garage are full of flammable substances and sources of ignition. All it takes is a moment of inattention, a bit of carelessness, a single instance of ignorance. Ask burn nurses how people end up in their care and you are likely to hear, "Whenever we have an empty bed, we know somebody somewhere is going to be striking a match sometime soon."

More often than not, that match is struck in and around the home, causing significant property damage and injury to other members of the household or residents of the building. Fortunately, everyone else in Dan O'Shea's townhouse got out safely, but his apartment was gutted and everything he owned destroyed. In the United States, half of all burns requiring hospitalization occur at home and, like Dan's, many are accompanied by inhalation injury. About 10 percent of serious burns happen in the workplace. The rest occur in public places or are related to recreational activities or motor vehicle accidents. The site of most residential fires is substandard housing, where living conditions may be crowded, smoke alarms and sprinklers scarce, heating and electrical systems faulty, building materials low-grade, and structures poorly maintained. The poor are on the list of people who most frequently get burned.

If many people are burned in their homes, many others who are homeless suffer from life-threatening burns, due largely to the risks involved in finding warmth and shelter outdoors. When a homeless burn victim is brought in to the General, identifying the patient and locating family members pose another challenge for the burn team. Most of it falls on Carla Cucinatti, the social worker whose colleagues refer to her as the Columbo of the burn service. An elegant woman with a girlish voice but a steely determination, she hardly resembles the bumbling television detective, but she never fails to crack the case, using all her skills and intuition to ferret out the facts. If there is no family and communication with the patient is impossible, she has to do a lot of digging. She may find someone in the community who knows or at least recognizes the patient, someone who works in a homeless shelter or an outreach program, other people who live on the street. Once she has a name, she tries to track down a family member, though it may be someone who lives far away and has been out of contact for a long time. Still, she cannot recall a single patient who remained an unknown in the five years she has been on the job.

Homelessness often goes hand in hand with mental illness and substance abuse, which, singly or in combination, account for many of the most severe burns. Sometimes they are a result of careless behavior and slow response while intoxicated, but increasingly people are turning up in hospitals with hand and face burns, including chemical burns as well as inhalation injuries, due to explosive circumstances associated with drug use and manufacture, from freebasing cocaine to compounding methamphetamine in a basement lab. As one burn nurse observed, "It's hard to make drugs when you're on drugs."

Most burns are accidental, but about 5 percent of all serious burns are intentional. It could be assault, arson, or abuse, but a lot of the time it is attempted suicide. On any given day on any specialized burn unit, there will be at least one "self-igniter," and often more. Of all the ways to end your life, self-immolation is probably the worst, if for no other reason than when you fail, you condemn yourself to hell on earth. There is no question these patients challenge everyone on the burn team. Self-inflicted burns tend to be extensive and life threatening. It is frustrating to work so hard to save the life of someone who has chosen to end it. One nurse voiced the frustration shared by many of her coworkers about these would-be suicides. "I wish we could go on national TV and say: If you want to kill yourself, don't set yourself on fire. It's the most painful death you can have, and it doesn't always work." Whether it works or not, she said, "It's traumatic for the family. People do it in front of their loved ones. It's always awful."

John Findley is the attending psychiatrist on the Mass General burn service, which makes him something of an authority on what leads people to try suicide by fire. According to Dr. Findley, they are not merely depressed. "Depressed people hang themselves, they jump off bridges, they cut their wrists. The ones who light themselves on fire are psychotic." The psychosis may coexist with major depression or schizophrenia or substance abuse, but for "whatever reason, they're not in reality, and setting themselves on fire becomes a solution for them."

He estimates that half of all burn patients have a preexisting psychiatric condition. "You could argue that we have more psychiatric illness in this unit than they do on a psych floor," he said.

In some ways this can be viewed as a blessing in disguise. The care and attention these very needy patients get on the burn unit can be healing emotionally as well as physically. "They are getting attention they never got before, despite how painful it all is. Here they've got a lot of people who are really invested in their care." Findley singled out the nurses and the hands-on care they provide, especially during the twice daily dressing changes. "That is something unique on this unit. The nurses really get to know their patients, they spend a lot of time with them, they're not just checking meds and then moving on to the next bed. So there's a real investment that the staff has for their patients, and that can be protective."

For some patients, the burn unit is their first encounter with that kind of caring. Carla Cucinatti told me, "Sometimes it's not until they're in the hospital that they get what they have been needing. It's like taking your car into the shop for an oil change and finding out there are all these other things that need fixing too. I see it as an opportunity. We have a captive audience and a chance to get things in place, or at least to plant a seed." Many burn patients leave the hospital with appropriate medication for psychiatric illness, referral to continued mental health services, the beginnings of recovery from substance abuse, help with housing, access to medical care, financial assistance, and many of the other strands of the safety net they desperately needed even before they were burned.

* * *

Two types of burns occur more rarely than flame, scald, and contact burns but can be more damaging. Electrical and chemical burns are most commonly workplace hazards. High-voltage electricity can kill instantly, but when it does not, it inflicts a burn that may be deceptive. A small burn at the point of entry masks extensive injury deep below the skin and into the muscle. The systemic effects of such burns are enormous, including the wholesale destruction of muscle tissue, a condition called rhabdomyolysis that makes catabolism look like kid stuff. The waste products flood the kidneys, producing dark, concentrated urine that one doctor described as having the color and consistency of crankcase oil. Prodigious amounts of resuscitation fluid are needed to clear the waste, but even then, the risk of kidney failure is high. Swelling deep below the surface wreaks havoc throughout the body. Removal of dead tissue down to the bone and, if the burn is in a limb, amputation may be the only ways to keep the damage from spreading.

Chemical burns from either acid or alkali are especially pernicious because the damage they do continues until they are neutralized. In the meantime, they cause tissue destruction by a variety of means, each of them ghastly. One of the most common chemical burns comes from contact with wet or even powdered cement. The culprit in this case is calcium hydroxide, $Ca(OH)_2$, a strong alkali that is produced when water or sweat is added to the mix. One of the most awful chemical burns comes from hydrofluoric acid, a highly corrosive substance used in glass etching. After burning its way through the skin, HF attaches itself to calcium, sucking it out of the bloodstream.

"You have to hack off big hunks of tissue, because the hydrofluoric acid takes the circulating calcium out and the heart stops," Bob Droste explained. He remembered a man burned by hydrofluoric acid who never made it up to Bigelow 13. "We had a guy with burned thighs go into the OR and they worked as fast as they could and they were dumping calcium into his blood as fast as they could and he died anyway."

* * *

The peak ages for getting burned are five to nineteen and thirty to thirty-nine. The incidence at the two ends of the age spectrum—from birth to two years old and sixty-five and older—hovers around 10 percent. The numbers may be much smaller than for those in the prime of life, but the consequences are far more dire.

The odds of surviving a major burn are a combination of burn size, the presence of inhalation injury, the existence of one or more other diseases or conditions, and age.

All other things being equal, patients between five and thirty years of age do best when it comes to survival, length of hospital stay, number and severity of complications, recovery, and rehabilitation. Burn victims under five and over sixty are more likely to die than those in any other age group; those who live spend more time in the hospital, suffer more complications, and are more significantly disabled by their injuries.

Even a relatively minor burn can be life-threatening in an elderly person. Exhibit A: An eighty-year-old woman is close to death after stepping into a hot bath. Diabetes has damaged the nerves and blood supply to her extremities, and she did not realize the water was too hot until she had eased herself into a sitting position. By then she had a superficial partial-thickness burn from her feet up the backs of both legs to her buttocks, and wound up in the hospital. They managed to pull her through burn shock without her heart or lungs giving out, but she is far from out of the woods. She is in an enormous amount of pain and is extremely disoriented. In a younger, healthier person, a burn like hers would heal in time, but diabetes interferes with healing by reducing the flow of oxygenated blood to the wound. Despite all efforts, gangrene has set in and doctors have little choice but to amputate both legs up to the knees. There is some question whether a patient that old and that sick can survive the surgery, and after that a lengthy hospitalization, with the risk of further infection, including pneumonia. Her kidneys are also in jeopardy—already damaged by her diabetes, they are being poisoned by her burns. If she lives long enough to leave the hospital, she will face extreme disability as a double amputee. The outlook for a patient like this is bleak.

Old people are hit with a double whammy. They are at risk for getting burned because of the decreased mobility and increased disability that comes with age. Physical, cognitive, and sensory (vision, hearing, smell, balance) limitations put them in harm's way and make it difficult for them to move quickly enough to escape. When they get burned, the injury tends to be more severe and their care more complex. Older people have thinner skin and a diminished fat layer, which makes for a deeper burn. They also have more health problems, a catalogue that includes cardiovascular and respiratory disorders, diabetes, and kidney disease just for starters, all of which can lead to complications despite the most careful monitoring. Diminished lung function makes them highly vulnerable to pulmonary edema, pneumonia, and respiratory collapse, all extremely grave complications. Their immune responses are weakened, making them prey to infections younger people can easily fight off. Poor nutrition, which is not uncommon in the elderly, is another complicating factor that has a huge impact on wound healing. Healing is also impeded by age-related changes in cell-repair mechanisms and impaired circulation. Their frailty makes them poor candidates for surgery, though surgery is what they often need. Decreased muscle mass hinders their ability to withstand the high metabolic demands of a burn wound and increases the length of recovery and the likelihood of long-term disability. Most elderly people take medicine for one and usually more chronic conditions; this requires skillful management to avoid dangerous interactions with the antibiotics, pain medication, and other drugs required for their care. Many older burn patients have memory and other cognitive deficits and little or no social support, further burdening their recovery and ability to cope during and after hospitalization.

For these and a host of other reasons, age is a major predictor of who will live and who will die from their burns. Even though she has only a 20 percent burn, the eighty-year-old woman is almost certain to die. What started as a small mistake may well end in tragedy.

A burn is sad at any age, but the saddest of all are the babies, whose lives, if they can be saved, are affected in more ways and for longer than any other burn survivors.

In terms of TBSA and depth, infants are at risk for the most severe burns. A splash of hot coffee measuring three or four square inches covers a much larger percentage of a baby's total body surface area than it does an older child's or an adult's; the size of a hot plate or radiator coil is the same whether it is burning a baby or an adult, but it burns a larger percentage of a baby's TBSA. Infant skin is also thinner, so the same amount of heat or the same length of exposure will result in a deeper burn. Being smaller complicates everything and narrows the margin of error every step of the way. Early on, the smaller the child, the greater the danger of heat loss to the point of hypothermia because of the greater surface area in relation to body mass, less insulating fat, and lower muscle mass, which generates heat through shivering. Because the fluid lost from their circulation after a burn represents a greater proportion of their total fluid volume than it does in adults, babies and small children are highly prone to shock, which can be swift and deadly. Fluid resuscitation is considerably more complicated. Accurate assessment of burn size and precise monitoring of vital signs are essential since the distance between hemodynamic stability and shock is much smaller. But estimating burn size is tricky with babies and young children, whose body proportions differ from those of adults and older children. Delivering the fluid is trickier still because their veins are smaller, requiring smaller catheters with less carrying capacity. Gaining access to their veins requires experience and expertise to avoid puncturing the lungs and other vital organs crowded into a tiny interior space.

Keeping babies breathing is crucial but also complicated by anatomy. Their tongues are proportionally larger but their airways are narrower and shorter, and therefore more likely to be blocked by swelling but less able to accommodate a breathing tube. Placing that tube, which should be no greater in diameter than the baby's pinkie, requires the utmost precision to avoid injuring delicate structures. Ventilator management must be overseen by someone highly experienced in pediatric respiratory issues. Settings on a mechanical ventilator have to be watched carefully and constantly, since it may be difficult to provide adequate ventilation through the narrow breathing tube, but higher pressures could damage a baby's small, delicate lungs.

That is the short version of why a young burn victim is in gravest danger during the first forty-eight hours, but the daunting treatment challenges persist long after the initial life-saving period. Pain management is fraught with difficulty: It is harder to assess the level of pain, medication options are more limited, and dosage calibration must be more precise. Burn-related metabolic changes that can be avoided or minimized in adults tend to be more extreme and last much longer in children, sometimes for a year or more after the burn. Vitamin deficiencies and low bone density are often seen in children who have been burned, resulting in nutritional problems and risk of fractures later in life. Severely burned children require long and often repeated hospitalizations, with multiple surgeries and extensive physical therapy and rehabilitation. When burns require grafting, there is much less healthy skin available and what is there is more fragile. Scars can severely restrict range of motion as a child grows, hampering development.

Burns related to child abuse are disturbingly common, with the youngest children being the major targets. The highest incidence of nonaccidental burns is in children between the ages of thirteen and twenty-four months. These range from cigarette burns and contact with hot radiators or irons and other electrical appliances to immersion in scalding water. Scalding, whether accidental or intentional, is by far the burn most frequently suffered by young children.

It does not take long for hot water to inflict a deep and extensive wound. Most water heaters are set at the factory to 140 degrees Fahrenheit, the middle of the heating range of 120 to 160 degrees. At 125 degrees Fahrenheit, water will inflict a second-degree burn in two minutes and a third-degree burn in four. It will take only three seconds for a second-degree burn and five seconds for a third-degree in water at 140 degrees. At 155 degrees, a deep burn takes only one second, faster than even the most agile person can jump out of a tub or shower. That is why it is recommended that water temperature be turned down to 120 degrees in households that include children, elderly people, or people with disabilities. Of course, this requires access to and control over the temperature gauge, something that may not be possible in an apartment building, for example. In that case, hot water should always be mixed with cold, and the temperature tested by a competent adult.

The kitchen is even more dangerous than the bathroom because there are more hot things, and those things are hotter. A freshly poured cup of coffee is typically about 180 degrees Fahrenheit, and boiling water is 212. Steam can be no cooler than 212 degrees, but it can be much hotter than that, depending on how much pressure it is under. Contact with steam released from a closed container like a pot with a lid on it, a teakettle, or a steam iron will instantly result in a burn. Steam from a pressure cooker can be as hot as 250 degrees, and from a steam pipe it can reach temperatures as high as 400 degrees. Cooking fats and oils start sizzling at about 350 degrees and burst into flames at 400. Sauces and soups may be no hotter than boiling water, but they are thicker, so they tend to retain their heat and stay in contact with the skin longer than a splash of water, which cools on contact.

Many scaldings are accidental, of course. Toddlers climb and reach and babies wriggle in laps. In many households, caregivers must choose between leaving them unattended and trying to keep an eye on them while cooking, cleaning, and ironing, among other distractions. Sometimes the line between accident and neglect is a fine one—as in the failure to properly supervise a child, maintain household water at 120 degrees, or in other ways childproof the home. But intentional scalding does happen, and burn centers see far too much of it. The federal Child Abuse Prevention and Treatment Act (CAPTA) requires medical personnel to report any instance of suspected child abuse. Fortunately, it is not difficult to identify. In the case of a scald, for instance, the absence of splash marks and burn margins that are clearly demarcated and identical on both sides of the body suggests the child was immersed in hot water.

One of the most obvious signs of a nonaccidental scald is a "red kneesocks" burn on both legs.

It is impossible to overstate the profound psychological effects a burn has on a young life. In addition to the trauma of the burn injury, a child may also be hit with the loss of one or more parents or other family members, home, and sense of security. All too often, the burn is the result of abuse, commonly at the hands of a parent or other trusted adult. Hospitalization may bring previously unreported abuse to the attention of authorities, but it will most likely be followed by sweeping changes in the child's home and living situation, not always for the better. Sadly, burned children may be tossed between frying pan and fire at a time when they need an enormous amount of support, stability, and skilled care.

* * *

It costs a lot to treat a major burn, upward of $5,000 a day, not counting surgery. At the General, an ICU bed alone costs nearly $4,500 a day, a semiprivate room for patients not requiring acute care is a bargain at around $1,600. Nursing is included in the price, but medications, oxygen, physical therapy, consults, special equipment, X rays and CT scans, and an assortment of bedside procedures are all extra. A single visit to the OR can run up to $40,000, and many burn patients have multiple surgeries. Physician services are billed separately by the private practice to which the surgeon belongs. The length of hospital stay varies from a single day to a year or longer. It is not unusual for a patient with a big burn and complicating factors to remain in the hospital for many months. So multiply that daily figure by 60, 90, 180; toss in three or four surgeries, a half dozen units of blood, X rays, pain meds, and three weeks on a respirator; if the patient survives, add to that a week or more in a rehabilitation facility and a year or two of follow-up visits to the outpatient clinic, and you are looking at a massive bill, well into the hundreds of thousands of dollars, possibly even the millions.

By and large, health insurers and managed care organizations foot that bill, *if* the patient happens to be insured. Well over a third of all burn center patients are not. When private insurance does not cover all or most of the cost, payment is cobbled together by an assortment of public and private funding sources. The government picks up a good chunk of it through Medicare, Worker's Compensation, and other programs, and, given the socioeconomic status of many burn victims, Medicaid, disability coverage, and other forms of public assistance. When uninsured patients come in to the General, every effort is made to enroll them in programs for which they are eligible. Many states, Massachusetts among them, have what is termed an "uncompensated care pool," which is paid into by state agencies, participating hospitals, and the private sector, and is set aside for state residents who are ineligible for any other funding. For patients who fall through all the cracks, hospitals swallow the cost of treatment as part of their free-care budgets. At the General, no one is denied care, regardless of ability to pay. This practice of noblesse oblige amounts to hospitalwide free-care expenditures topping $20 million annually.

No one is getting rich off treating burns, a field of medicine regarded by many as an awful lot of hard work for an unworthy, if not lost, cause. John Schulz says it flat out: "There are people who think what we do is a sin. Saving burn patients. There are definitely people out there who think that people who get burned deserve what they get and why should we spend all these resources on them. Obviously, we don't agree."

Burn care is time- and labor-intensive, with a budget-busting staff-to-patient ratio. Twice-daily dressing changes leave a mountain of rubber gloves and paper masks and a trail of gauze that could wrap the planet. If there is no shortage of burn patients, there is a grave shortage of people to treat them. The demand for burn surgeons far exceeds the supply, and the pipeline is an echoing void. Burn surgeons fall at the low end of the income scale in a specialty that pays its practitioners handsomely. Mass General Burn Associates have been known to waive their fee when a patient has no means to pay it. This is undoubtedly the case in other burn centers as well.

Most burn care professionals acknowledge that no one—neither family nor friends—wants to hear about their work. Between the sights and the smells, this is no place for the squeamish. Clearly, something other than money and prestige motivates this community of healers. Some say they like the intensity of the work, some talk about the teamwork and unusual degree of camaraderie. Some value the deep connection they make with patients and their families. Some point to the challenge of saving lives at risk and the satisfaction when they succeed. No matter whom I asked, I always heard some version of this: "When there are bad times and there's a lot of death and a lot of sadness, all it takes is for one person to come into clinic who's doing well to make you say: This is worthwhile."

* * *

There is a saying in the burn treatment world that "Normal people don't get burned," and it is only half gallows humor. Statistically, the people who fill the nation's burn units are among the poorest and least visible: the homeless, the mentally ill, the would-be suicides, the addicted, the aged, and the abused. But consider the case of Tom Parent, the archetypal normal guy—solid citizen, homeowner, weekend sailor, loving husband and father.

Tom is a tall, husky man who looks like a high school athlete gone slightly soft in the twenty or so years since he last took the field. His wiry black hair has a bit of gray in it, but he exhibits a youthful physicality. Or maybe it is the incongruity of his mouthful of braces, a middle-aged adventure in orthodontia. He is a TV news cameraman for an ABC affiliate in southern Connecticut, and it is not hard to picture him standing out above a crowd, video camera on his shoulder, jockeying for position or running alongside the action to get his shot. He might have been voted Class Clown or Most Outgoing for his high school yearbook. He is the kind of person who makes friends wherever he goes.

It was a Memorial Day weekend at the Parents' vacation house on Cape Cod. Cooking oil in a pot on the stove burst into flames, starting a chain reaction that landed Tom in the Mass General burn unit and changed his family's life forever. What happened that day and in the months that followed happened to Tom's wife, Nancy, as much as it did to Tom. She has not been able to let go of the conviction that it was all her fault and she wears her guilt as openly as Tom wears his scars. Except Tom can joke about it, pointing out that the scar near his left thumb is shaped like the state of Texas and shaking his head in fatherly pride that the only one in the family who knew what to do was their teenage daughter, Ashley, while the grown-ups stumbled around like a couple of Keystone Cops in a surrealist's nightmare.

The way Nancy tells it, on May 26, 2001, they did everything wrong. But the truth is that what happened was a horrible accident, the kind that could befall anyone. It resulted from a series of miscalculations that added up to disaster.

Tom, Nancy, Ashley, and Nancy's seventy-nine-year-old mother had driven out to the Cape from their home in Ludlow, in western Massachusetts, for the holiday weekend.

Though traditionally the beginning of the summer season, that particular weekend was a chilly one. It was around two o'clock on Saturday afternoon, and Nancy was in the kitchen trying out a new recipe that involved flash frying chicken in peanut oil. A pot with six cups of oil was heating on a back burner while Nancy chopped onions on the counter to the right of the electric stove.

Just about that time Tom came in from working on his boat and went into the bedroom to exchange his nylon windbreaker for a sweatshirt. Nancy's mother was in the living room, and Ashley was on the phone in her room when she heard her mother scream.

Tom heard the scream too and rushed into the kitchen, with Nancy's mother at his heels. Flames were leaping from the pot, licking the underside of the microwave, which was mounted above the stove. The lid of the pot sat on the counter near where Nancy had been working and the sink was just inches away, but Nancy was paralyzed and Tom was focused on the curtains that framed the window behind the sink and hung perilously close to the route from stove to sink.

Tom was thinking fast, and he thought he was thinking clearly. "I've dealt with fire before. Heat rises; everything below the fire is cool. The handles are below the fire. Well, this will be no problem. I'll just take it from the stove and out to the patio."

Picking up the pot by the handles and pivoting toward the door put Tom directly under the ceiling fan, its blades turning lazily a foot or so above his head. Nancy's eyes followed him in horror as the flames were literally fanned to about double in height. She stood rooted to the spot as she heard Tom bellow, "Fire!" and watched him head for the family room, which led to a screen porch and the backyard beyond. "By the time he got to the doorway, I couldn't see him from the waist up. I couldn't see him at all. I could only see his legs." The rest of him was engulfed in flames.

The moment she heard the word *fire*, Ashley made for the front door and ran to the neighbors' house. She did as she had been told a dozen times. She just ran and did not look back to see her father burning.

Nancy was not spared the sight, and she will never forget it. "The dog was lying in front of the screen door, and I knew that's where he was heading, but he couldn't get there because she was there, so he went towards the couch, and all I could think was, 'The couch is going to catch on fire, the dog's going to catch on fire, and I'm not going to be able to see any of them.'"

Finally, Nancy thought about the fire extinguisher, which she knew was in the front closet, except it was actually in a supply closet in the kitchen, but she did not know how to use it anyway. As Nancy says, that was a bad time to stop and read the directions.

Tom was still running on adrenaline when he realized his route was blocked by the dog, a Samoyed with hair as long and fluffy as a sheepskin rug. That is when he looked down at his hands and saw the skin curling off his fingers in a molten sheet. He hurled the pot at the screen door, dropped to the ground, and rolled. Seconds later, he bounced up again, steadying himself against the couch as he rose, holding his blackened hands out in front of him. The couch is still there, but Nancy keeps the middle cushion turned over to hide the ghostly print of a hand scorched into the slipcover.

Nancy experienced the whole thing in slow motion. It felt like forty-five minutes had elapsed, though it was more like forty-five seconds, from kitchen to couch. After it was over, she remembers thinking, "This isn't too bad." She looked at his hands and his forearms, bare to the elbows where he had pushed the sleeves of his sweatshirt. They looked gray, and she figured he was okay. He was talking; he was standing up; he seemed to be all right. He was saying, "Get my watch off," so she started to do that when he spoke again. "Get me to the hospital."

Nancy was working on autopilot. "Get me to the hospital" meant *Get in the car and drive to Cape Cod Hospital*, just a few miles down Route 28 from their home. It did not mean *Call 911*. And it did not involve remembering that Route 28, the main artery that runs from the Bourne Bridge and Woods Hole along the Nantucket Sound, through Hyannis where the ferries leave for Nantucket and Martha's Vineyard, past miniature golf courses and stands selling soft ice cream and fried clams, skirting marinas and motels, and winding its way through West Yarmouth and the handful of other towns that run up to the elbow of Cape Cod, is a bumper-to-bumper ribbon of cars from Memorial Day weekend to Labor Day.

By the time they got to the top of their street, the folly of it had dawned on Nancy. She looked at the traffic-clogged road and then she looked at Tom. He was rocking back and forth, still holding his arms out in front of him. "I can't make it, I can't make it. Go to the police station." The police station sat directly in front of them, just across Route 28.

Nancy talks about what happened next as though it were a miracle, her voice thickening at the memory. "No one was in the intersection. All the cars at the intersection had stopped and I just sped right through. I swear there were angels at every corner stopping those cars."

By then, the pain had become unbearable. Maybe it was part of Tom's strong survival instinct, but he seemed to disassociate himself from what was happening to his body. There he was, inside the police station, conversing coolly about the department's new canine unit with the officers who were trying to tend to him. It seemed to Nancy that the police did not know any more than she did about what to do. It was Tom who took command. He asked where the men's room was and said, "I've got to get my hands underwater." That lessened the pain a bit and probably delayed, but could not prevent, the swelling. Then he told them to cut off his sweatshirt, which had withstood the flames and protected much of his upper body. When they balked, he insisted, barking out orders like it was all happening to someone else. And Nancy just kept telling herself, "This is fine. He's fine. Everything is going to be okay."

The EMTs arrived about five minutes later. They recognized immediately that Tom was not fine, that he was heading into shock. His skin was pale and clammy, his breathing was rapid but shallow, his heart was racing but his pulse was weak—"thready" is the expression medics use. The assertiveness that Nancy took to be a sign he was okay, they read as the characteristic excitability that precedes full-blown shock. One of them got the hospital on the line and relayed to his partner the instructions he received: IV with normal saline, a squirt of morphine, and dry, sterile bandages everywhere Tom's skin was blistered, charred, or melted. They got him into the ambulance and wrapped him in a blanket, but it was another fifteen minutes before they headed down the road to the hospital. That gave Nancy time to return home to check on Ashley and her mother, to let them know what was happening, and to grab Tom's cell phone.

Nancy followed the ambulance to Cape Cod Hospital, but by the time she parked and got inside a nurse told her Tom was being transferred to the General. "Look, we've taken some seminars," she was told, "but I don't think you want people who have just taken seminars dealing with this. His hands are burned, one of them pretty badly, so we're going to send him to Boston, where he ought to be."

It seems counterintuitive, but burn victims are at their most transportable right after they have been burned. If they can be kept from going into shock, there is a twenty-four-hour window, plenty of time to get them to a specialized burn center where they can get the care they need. Tom Parent was lucky the doctors at Cape Cod Hospital knew that. Knowing when you do not know enough is almost as important as knowing enough, in burns as in life. They knew they did not know enough to treat a major burn. They also knew when a patient should be sent to a specialized burn unit, and they could see Tom qualified.

Tom's hands were severely burned. One side of his face was very red, especially around his mouth and eye, and his eyebrow and hair were singed on that side as well, all of which raised the possibility of inhalation injury in addition to the obvious face burn. A lot of the burns looked second degree, but the doctor who examined Tom at Cape Cod Hospital knew it was hard to tell just by looking. Tom is a big man, but between his face, his hands and arms up to his elbows, and some burns around his waist, it could all add up to 10 percent of his total body surface. On top of all that, they had found out when they took his history that he had high blood pressure and was taking medication to control it. It was just not worth the gamble, especially since he was conscious, alert, and stable and could be kept that way with an oxygen mask and IV fluids during the two-hour trip to the General.

Tom was more than conscious and alert; he was positively wired. Nancy was not in such great shape herself. "I was beyond hysterical," she admits. While Tom was being packed up for the journey to

Boston, one of the nurses made Nancy promise she would go home and stay there until she was calm enough to drive into the city. She welcomed the order because she wanted to spend some time with Ashley and her mother, to be sure they were not traumatized by what had happened. She also needed to arrange for her brother to eventually come get them and drive them home to Ludlow. She managed to get herself in check enough to accomplish those tasks and pack some clothes for herself, but *calm* is probably not the word to describe her. "I may have been calmer, but I don't think I was calm until about thirteen months later."

Meanwhile, Tom was almost enjoying himself in the ambulance, though he admits he was pretty "zoney," thanks to his surging adrenaline singing an off-key duet with morphine on demand. He was lying on his back, his arms suspended in Ace bandage slings from an overhead railing, to keep them above his heart and enlist gravity to stem the tide of fluid surging toward his burns. They listened to a Red Sox–Toronto game on the radio and Tom did his usual thing, getting to know everyone's life story within minutes of meeting them. "One of the guys graduated from Mass Maritime Academy. He'd been working on ships for a long time, and we're talking about his job as a seaman and stuff like that. It was a great crew all the way up." He managed to make friends in the Emergency Department as well. By the time Nancy arrived at the General around six that evening, Tom was sitting in one of the recliners in the major multipurpose area, telling jokes to the scrub-clad assembly. One of the nurses turned to Nancy and said, "We're not going to send him up to the burn unit, we're keeping him here. He's just too funny." That is when Nancy caught sight of Tom's left arm.

"I couldn't believe it. His arm looked like a cartoon arm, his hand was literally four times as big as it really is, puffed up like Mickey Mouse. It just took my breath away." Tom's hand had already begun to swell back on the Cape. Charred hot dogs—that is the image he cannot shake when he thinks about the fingers of his left hand. They had tried to get his wedding ring off in the ambulance and again in the ER at Cape Cod Hospital because they were afraid it would strangle his finger, and that could mean amputation, but it was already too tight to slip or twist off. Tom fought them about cutting it off, but Nancy knew it was no time to be sentimental. "Just cut the damn thing off!" she ordered.

If Tom's role is to provide comic relief, Nancy is the take-charge center of gravity that keeps their family going. She is smart and articulate, earnest and honest about feelings and events. A generously built woman, she gives the impression of being both soft and strong. A hug from her, you can be quite certain, would make you feel safely pillowed against harm. And if you needed someone to speak up for you, you could not find a better advocate. Nancy knew Tom would be going upstairs to Bigelow 13; she knew this was more than an overnight affair; and she had every intention of being at her husband's side throughout whatever lay ahead.

CHAPTER FIVE

Bigelow 13

When you take the elevator up to the thirteenth floor of the Bigelow building, you enter a world most people, if they are lucky, have no idea exists. A high semicircular reception desk stands between two corridors. The one on the left belongs to the plastic and reconstructive surgery service, a separate realm ruled by an entirely different group of physicians. The right-hand corridor is the domain of the Sumner Redstone Burn Center. At the top of the corridor is room 1322, a 1,100-square-foot rectangular open space that is the beeping heartbeat of the floor. This is the burn intensive care unit. For some reason, it is exempt from the usual hospital alphabet soup—MICU, PICU, SICU—and is simply called "the unit." Here the sickest burn patients lie inside clear plastic tents, each with its own tropical microclimate to keep the occupant warm and protected, as much as possible, from infection. There are four bacteria-controlled nursing units—BCNUs, as the tents are called—and it is rare to find one of them unoccupied.

Strictly speaking, the entire right side of the floor is the burn unit, with single patient rooms on the window side of the corridor numbered from 1324 to 1330, ending with two examination rooms and an alcove that serve as the outpatient clinic. As is true at many hospitals, the patients who need the most attention are closest to the front; the farther down the hall patients get, the closer they are to going home. They say that a burn bigger than 20 percent buys you a bed in a BCNU. Smaller burns or big burns that are well on the way to healing go to room 24, just outside the unit. Patients in this room tend to need critical care—ventilator, one-on-one nursing, close monitoring—but not the special environment of a BCNU. Next come three general care rooms for what are called floor patients, who no longer or never did require intensive care. If there are more patients than available beds, the burn service borrows from the plastics side.

That was the case the night Tom Parent arrived. The burn side was filled to capacity so Tom was put into a double room on plastics, because, as Tom said, "I was the least worse." He was still the responsibility of the burn service, and he would be cared for by nurses with burn experience. The room assignment was strictly a matter of real estate.

It was well into the evening shift by the time Tom got up to Bigelow 13 and was installed in room 1342. The nurse assigned to him that first night was Frank Ireland. Nancy stood by feeling useless as Frank helped Tom into bed, adjusting the angle of the bed so his head was elevated and gently but skillfully positioning his arms, which had been splinted to keep them immobile with his elbows bent and hands well above his heart.

Frank is a teddy-bearish man, with a shy smile and a calm manner. He came to Bigelow 13 in the mid-1980s as an operations associate—front desk clerk—but was grabbed by whatever it is that grabs people who end up as burn nurses. He met his wife, Lisa, on the job and they now work alternate shifts so one of them is either nursing burn patients or caring for their three kids. Presumably they get to sleep at some point, though it is hard to imagine when. People talk about the burn team as a family, but for Frank and Lisa it is literally true, and apparently with the hospital's blessing. "After we got married, we weren't sure they would let us both work here. I don't know what would have happened if they said no, because I don't know which one of us would have left. We both love being here, absolutely love it." Not only is it allowed; the family way seems to be part of the culture at the General. Two other nurses on Bigelow 13 are married to nurses on other floors, and Bob Droste probably holds the record. His wife and two of his children are MGH employees, though he is the only nurse in the family.

Frank clicked a vial of morphine into place and showed Tom how to operate the patient-controlled analgesia pump. He hung up a bag of Ringer's, set the infusion rate on the pump, and opened the stopcock leading to the line in the crook of Tom's elbow. Frank had introduced himself so Nancy knew he was the nurse, but she asked to talk to a doctor. She was still in a state of high anxiety, and she had a lot of questions about what would come next.

Less than a minute later, a young woman in a white coat, a stethoscope draped around her neck, walked into the room. This was the doctor, the resident on call that night. Nancy does not remember her name, and it is unlikely any of the nurses would be able to tell you who exactly was rotating through burns in May 2001. Nancy felt as though she and Tom had walked through the looking glass. In this inverted world, the nurse was a man, the doctor was a woman, and it would be the nurses and their aides, more than the doctors, that both Tom and Nancy would come to rely on throughout his stay. That night, Tom was pretty much out of it, but soon he came to understand the lay of the land. "It took me awhile to figure out how the system worked. I knew that the A team was Schulz, Sheridan, and Ryan. And then there were my buddies—the nurses and the techs—and they were the ones who were there all the time, who made sure I was comfortable, who let me know what was going on, *who took care of me.* And then there were these people that would come in and I had no clue who they were, but they must have been the residents."

* * *

More than 700 residents rotate through the General in any given year, in one of twenty-six specialties. The length of the residency varies, though none is less than two years. The surgical residency lasts five years. Some residents add a year or more of research, either after completing clinical training or as a break from it. For surgical residents, the clinical years are spent on a merry-go-round in the wide world of general surgery: a total of eleven months on the surgical teaching ward and the rest a series of two-month rotations on twenty-one other surgical services. A resident intending to become a cardiac surgeon, for example, will spend time practicing everything from surgical gynecology and oncology to trauma and urologic surgeries—the whole gamut of the cutting arts. During those first five years after they graduate from medical school and pass their licensing examinations, residents are the principal hands-on doctors for the more than 40,000 patients who fill the General's 868 beds each year, as well as the million plus who are cared for in the Emergency Department and outpatient clinics.

Although it may be true that the latest and presumably the best medical knowledge is available at large teaching hospitals, the other side of the coin leaves an ever-changing cast of sleep-deprived physicians-in-training writing the orders and wielding the scalpels. This state of affairs breeds confusion and distrust in many patients and their families, at the General as at all academic medical centers: "Which ones are the doctors, and why are they so young?"

It is also the source of no small amount of tension with the nursing staff, whose permanent presence at the bedside and extensive experience make them the patient care experts, the superiority complex of some of the new MDs notwithstanding. There is a lot of eye rolling among bedside nurses on the subject of residents who do not know how to write orders, fumble their way through procedures, and ignore or misinterpret signs and symptoms that an experienced physician, and the nurses themselves, would not. As Mike Wilson, who has worked as a nurse in the burn unit for years, put it, "When a young resident is trying to put in a central line, and it's the fifth or sixth stick, and you say, 'You know, perhaps we should take a break,' or 'If that was your mother, how many sticks would you think would be legitimate?'"

Mike is the gadfly of Bigelow 13, the keeper of the humanistic flame, the preacher of the gospel of holistic care. If it were not for his impish manner, he might not be able to get away with his outspokenness. "There are times when procedures are done that I know are being done for the learning effect only, and I have to ask: Where is the ethics involved in that? How far does a teaching hospital teach as opposed to when does it become undignified for that patient? The nurses are the only go-between in that. To intervene in that way takes an experienced nurse, and I know that we're considered obstacles in ways."

John Schulz vigorously disputes Mike's assertion. "We do not do procedures without a clinical indication. Who does the procedure is another matter. When residents do procedures it is so they can learn, but medically unnecessary procedures are never done for educational, or any other, purposes."

It is an uneasy alliance, at best. As patient advocates and patient care experts, nurses constantly have to weigh politics against what they judge to be a patient's best interests. They also have to practice diplomacy when bridging the knowledge gap that yawns between residents and themselves. Knowing how and when to question a resident's decision is a requisite part of every nurse's skill set. Mary-Liz Bilodeau is both a nurse practitioner and clinical nurse specialist, a set of advanced-practice nursing credentials that puts her somewhere between the bedside nurses and the medical staff, a position that requires political acumen, which she has in spades. The way it works, she said, is "If a nurse decides that this resident isn't doing what they need to be doing, they're going to call the attending and say, 'You know, I'm not really happy with what's going on. Maybe you might want to take a stroll by.' So the attending kind of knows that that's the red flag that they better get their . . . um, selves down here."

John Schulz himself is a product of the MGH residency program, so it is perhaps not surprising that he sees it in a more positive light. He does not view a resident's initial inexperience with burns per se as a problem. They are, after all, surgical residents and, he said, "Most of them are pretty interested in critical care—taking care of sick people. They're interested in learning operative techniques, so even those who are interested in being GI surgeons or cardiac surgeons in the end need to have a very fundamental grasp of how to take care of someone who's very ill. That's one thing that's uniform across the subspecialties in general surgery, and so within that kind of framework, burns fit not too badly." He is a vocal champion of residents, and his residents in particular. "One of the most notable characteristics of most of them is the high level of interest and commitment they bring to the job."

On the burn service, each pair of medical trainees consists of a junior and senior member, usually a first-year resident—what used to be called an intern—and a third-year. Classically, residents are the serfs in the feudal realm of hospital medicine. They are considered the house staff and are hospital employees, whereas the attendings are independent contractors. The term *resident* hints of the hours they log. Throughout their training, they practically *live* at the hospital, and are on duty most of that time.

The schedule on the burn service puts the junior resident on call every third night and the senior one night out of every seven. On-call residents bunk in a suite behind the door marked 1336, at the end of the burns corridor opposite the outpatient clinic. Two tiny bedrooms that are a cross between a college dorm room and a railway sleeping car share a bathroom and a revolving cast of characters. A narrow bed, a desk, and a five-inch television on an extendable arm account for the furnishings.

Not that it matters, since a resident on call rarely has a chance to sit down, let alone get horizontal.

On-call nights would not be so bad if they did not fall between two full days. Jennifer Verbesey, who put in time as a junior resident on the burn unit, remembered it almost fondly. "Actually there's some fun part of call. Almost every night they would order dinner so there's camaraderie." After dinner, she would stay around for a few hours, until she was sure things were calm. Some nights she was able to log four, maybe five hours of sleep. "I would usually go to bed around midnight, and then be up around 4:45. The nurses there are excellent and won't bother you for small things until the morning."

The overnight is an entirely different world, populated by the patients and the nurses and practically no one else. A respiratory therapist is on call, as is the resident sleeping at the end of the corridor and, if need be, the attendings, and there is someone at the front desk to answer the phone. But the hurly-burly of the day is absent. There are no rounds, no physical or occupational therapy sessions, no visitors, no business. In the unit, the second dressing change of the day takes place in the late afternoon or early evening, before the patient care techs leave for the day. Floor patients, whose dressings are much less complex, may get theirs changed after dinner. By the time eleven o'clock comes around, the floor is quiet and what activity there is centers in the unit.

The nurses are at their stations outside the BCNUs, checking vital signs, suctioning airways, keeping the fluids flowing, the meds dripping, watching the monitors, listening for any change in the steady beeping syncopation.

"The big thing that keeps you awake at night is a very sick patient," Jen said, "and also if you have an admission come in." The on-call resident is notified when a burn patient is expected down in Emergency. The call might come when a patient is transported straight from a fire, as Dan O'Shea was, and the ambulance is just minutes away. The General gets a lot of transfers, patients like Tom Parent who go to the nearest hospital but whose burns are too serious to be handled there.

"You usually know pretty far in advance. Many times I knew over an hour in advance that a patient was going to come in. Your beeper will start going off, you'll talk to the doctor at the other hospital, if the patient's being transferred from somewhere, which was frequent, because most places don't have burn units, or if it's an ambulance that calls in. And then when they come in, they beep you and you go down to the emergency room."

Burns do not happen on schedule, and as everyone knows, bad things tend to happen at the worst possible time. Dealing with someone fresh out of the fire is far from a solo act. "There are senior residents in the Emergency Room that can help, a senior surgical resident who could always peek in to say 'Yes, that's the right plan.' If it's a very, very bad burn, then your senior resident would come in from home—because they're always on second call with you. If it's something that they felt you could deal with, you'd discuss it on the phone. And then you would call the attending also. So mid- to moderate-size burns, after you'd been there awhile and were comfortable, you could handle on your own. Anything much larger or anything requiring an intervention, like an escharotomy, then the senior resident and the attending would come in."

Escharotomies are emergency surgeries. There is no time to lose, so they are usually done in the Emergency Department or in the unit if the patient can be moved up there quickly, not in the OR. They are not something a resident would or could do without supervision and prior consultation with an attending.

When she was not on call, Jen would wake up at 4:00 A.M. at her home in the Boston suburbs. "My rule has always been, I will not get up before four o'clock. I will get up at four, that's fine, but not before then, just for survival sake." She described a breathless morning jog around the floor. "You basically hit the door running. I'd get there early, about quarter to five, because my responsibility was to preround on all the patients. That entailed getting all the vital information on the patients for the last twenty-four hours, speaking to the nurse, seeing what happened overnight, examining each patient, and writing a note. As you can imagine, when you have five ICU patients, it takes a long time. And then I did the floor patients too, and then sometimes we had patients that were on other floors as well."

"Part of me enjoyed being the person who pretty much knew everything going on with every single patient. If it wasn't at five o'clock in the morning I would have enjoyed it more." The senior resident would arrive around six o'clock and the two of them would sit down at the oval conference table at the back of the unit that serves as office, break room, dining room, and town square, often all at the same time. By the time the residents get there, the table looks like an unmade bed and the overnight nurses are looking like they need to go home. Having tended their patients through the darkness and reported to the junior resident, what remains for this ragged band is the handoff to the day nursing shift.

Jen and her senior would go over all the information she had gathered, and then together they would make a plan for the day. "Different teams come up with different ways of doing things, but at Mass General all the orders are put into the computer, so he [the senior resident] would make some quick rounds and see the people again, and I would enter all the orders we had discussed into the computer, and then we would usually finish both around the same time."

"And then," she said, "the day would begin."

If it is a Monday, Wednesday, or Friday, it is surgery day. Dr. Ryan takes Mondays, Schulz Wednesdays, and Sheridan Fridays. The residents take them all. As close to 7:45 as possible, Jen and her senior would go down to the OR. "We pretty much operated all day. And in between, we would run back to the floor, take care of certain errands." That included discharging patients. "If anyone's going home, it's the junior resident's responsibility to get the discharge summary done, all the paperwork, and get the person ready to go."

Another job that falls to the junior is changing lines. "All the central and arterial lines get routinely changed. I would usually run up between cases to do that. Sometimes that takes five minutes, sometimes it takes an hour, and lines certainly are not easy with those patients." Changing a line is part of a physician's necessary skills, but finding veins and arteries can be tricky in a burn patient's swollen body. Detecting a pulse through burned flesh is close to impossible without a Doppler ultrasonic flowmeter, a wandlike instrument that can pick up the waveform made by blood moving through the arteries. The hypotension—low blood pressure—common in critically ill burn patients means blood vessels may be collapsed and hard to access. When a patient has a big burn, it is also difficult to find a clean, unburned patch of skin into which to place a line.

Having done her SICU rotation before coming to burns, Jen was initially shocked by the contrast. "In nonburn patients, you'd put lines in the most pristine areas, but here, look where we're putting in the line—

into raw, open skin. But when someone's whole body's like that, we really have no choice."

How frequently lines are changed varies from hospital to hospital; the range tends to be from three to seven days. At MGH, lines are changed every four days. That gives residents a lot of opportunity to get it right or wrong. Mike Wilson may experience each needle stick from the patient's point of view, but for Jen, "it's a great experience in terms of increasing your technical ability. I don't think there's anybody I can't put a line into now." Over the course of her time as a junior, she figures she put in more than one hundred lines. It is one of the many things a doctor needs to know how to do. Practice makes perfect.

All of the attendings—Colleen Ryan and Rob Sheridan as well as John Schulz—work very closely with the residents. Much of the teaching happens in the OR, of course, but a lot goes on during morning rounds, a ritual that can be humiliating or enlightening, and often is both. In the grand Socratic tradition, the attending shoots questions at the residents, asking for explanations and justifications for decisions made, probing more deeply, pointing out errors, posing what-if scenarios, and imparting bits of wisdom. Each attending has an individual style, making rounds a game of Russian roulette. At their worst, rounds bring to mind such expressions as "grilling" and "raking over the coals"; at their best, residents gain the expertise they need to think on their feet. Burn rounds at MGH are at the "best" end of the spectrum—constructive and supportive. All three attendings are superb teachers, less interested in formality than in sharing their knowledge. Whatever the style, for the outsider watching this interaction, it is easy to forget that it is really about the patients, not the residents.

Rounds take place when an attending is around, which might be first thing in the morning or later in the day. On surgery days, Jen and her senior usually met up with the attending down in the OR and rounds would take place between cases. "We'd just fit it into the day somewhere. The turnover time for the operating room gives you a half hour, forty-five minutes, so we'd say, 'Let's round after this case.'" Then the trio would go upstairs while the OR was being cleaned and prepped for the next patient. "On clinic days, we usually did it right after clinic. We always found a time. We'd try to go to each bed, where we'd present everything that we'd found out in the morning. We'd have the nurse from that person come to the bedside and be with us. They always have a lot of information to share."

The way the drill usually works is the senior resident gives the attending an oral report on each patient while the junior resident listens. "So basically the junior resident does all the work of gathering the information, the junior resident tells the senior resident, and the senior resident presents it all to the attending." Thus does the hierarchy of medicine extend down into the resident ranks. The senior reports to the attending, but he or she assigns, oversees, and instructs the junior in procedures and tasks, routine and otherwise. To quote one of those aphorisms that medical insiders find amusing but the rest of us chilling: "See one, do one, teach one."

Of course, attendings pay a lot of attention, even though they stand back to allow residents to gain the experience they need. Attendings do not rely solely on the residents' report. They talk to the nurses and independently check patients' results on the computer. An attending physician must be present in the OR, whether wielding the knife or watching. At its best, it is a hands-on tutorial and a group activity. "I try to do everything with residents," Schulz said. "The challenge as a surgical and a medical educator is you gotta *be* there, but you have to try to be there in a way that's not too obnoxious to the person learning. If you have good senior residents, they're doing most of the stuff, but you're always there, to back them up."

In addition to mentoring a revolving cast of surgical residents on both the burn and trauma services, Schulz teaches Patient/Doctor 1, a required seminar for first-year students at the medical school on the doctor–patient relationship. The idea, he explained, is to get them to talk to people like people. "Some of them come in without that talent."

* * *

The resident talking to Nancy Parent that first night either came with the talent or learned it under Schulz's tutelage. She spoke directly and told Nancy everything she needed to know right then, and in language that was neither too complicated nor too simplistic for an anxious but clearly intelligent family member to absorb.

"Let me start the conversation with: He's going to be fine." Nancy remembers letting go of one layer of tension when she heard those words.

Tom's life was not in danger. "He is in the most amount of pain he can be in because he's got second-degree burns. But that's also good because it means his nerves are intact. I know that's hard for you to imagine, but you're going to have to hang onto that for the next few weeks: The pain is a good thing. We'll do everything we can to keep his pain manageable, but we're never really going to be able to keep him pain free, so you're going to have to help him."

The young doctor explained what second degree meant, but also said it always takes a few days for burns to "declare" themselves—some might be deeper than they looked, some might require surgery in order to heal in the best way. Tom was lucky that only a bit more than 5 percent of his body was burned. That meant he was not at risk for many of the worst complications of a serious burn. And if he did need skin grafts, there would be plenty of healthy skin to "harvest." The burns to his face were first degree and therefore were not included in the TBSA, but they were a red flag because even a superficial facial burn meant exposure to smoke and flame. So the doctor was concerned about smoke inhalation and possible damage to his left eye. An ophthalmologist would come in to check Tom's eye in the morning, and his respiratory status would be monitored closely. The greatest danger, the resident explained, was infection.

He was going to be all right. Nancy listened carefully to everything the doctor said, but she was most grateful for the take-home message, or at least the message that would make it possible for her to leave Tom that night and try to get some sleep in a nearby motel, where a social worker had reserved a room for her. "That is a tremendous gift when you're in that situation. So as horrific as the day had been and as worried as I was about the next few weeks, at least I could go away that night and know that it wasn't life threatening, and that he was probably in the best place in the world, definitely in the country, but maybe the world."

* * *

There was no burn unit at the General after the Cocoanut Grove fire. The isolation wards set up for the victims were dismantled and returned to service as general-purpose floors. The nursing staff shrank to normal, the surgical staff went back to treating all manner of surgical patients.

Burn victims became the patients of whichever surgeon happened to be on call when they came in, but there was no burn service per se. The burn victims who did come to the hospital for treatment came in ones and twos, from house fires and other smaller disasters, with the occasional industrial or workplace accident swelling the ranks, though never on the scale of that November night in 1942.

Burn patients were housed on the top floor of the White building, designated White 12 despite the fact that it was actually on the thirteenth floor. Some say it was superstition, others claim it was because the entrance to the third-floor operating theater was designated floor 3A, throwing off the numbering of every floor above that. White 12 was a general isolation floor for patients who were vulnerable to infection regardless of what injury or illness brought them there.

Bob Droste took care of burn patients on White 12, one of only two nurses still on the job who remember the bad old days. Lynn Bellavia arrived in 1970, three years before Bob, but since she took time off for maternity leave, Bob can claim the longest tenure.

Everyone who worked on White 12 talks about the screen doors. That includes Bob and Lynn and even John Burke, the beloved medical director who brought the burn unit to Bigelow 13 and into the modern age. They were real summer house, bang-shut screen doors. "We used to take rubber tubing and tape it in the doorjamb so when the door closed it wouldn't slam," Bob told me.

The screen doors, Dr. Burke explained, were meant to keep out flies. "Flies were always considered bad things. All manner of infectious diseases were supposed to be put there—medical patients, kids, everything. And because burns have all kinds of infection problems, burns were up there too. But there wasn't any special facility organized for burns. The special facilities were organized to prevent infections from being spread."

By the time Bob got there, the floor was divided in two, one side for burns and the other for dialysis and kidney transplant. "It was a pretty decrepit, ancient place. It hadn't been renovated really ever. There was no air conditioning or ventilation. The windows had to be opened for ventilation. Before we left there we had gotten rid of the screen doors."

There is a portrait of John Burke on Bigelow 13, his handsome face with its aquiline features gazing benignly at the reception desk. It is a fitting position for the man who took the burn team from screen doors to bacteria-controlled nursing units. It was he who designed the BCNUs and built the burn intensive care unit around them. Nowadays you will find the man himself in an office clear across the MGH complex, in the far corner of the ninth floor of the Warren building, which is itself a warren of small offices apparently unrelated to each other. His suite is nicely furnished, the walls lined floor to ceiling with dark wooden bookcases, shelf after shelf of leather-bound and gold-stamped spines presenting a dignified facade. This is the domain of a professor and chief emeritus, still a name to reckon with but no longer involved in the daily activities of the hospital that was for so long his home.

Burke is above all a marvelous raconteur. It becomes clear as he speaks that the arc of his professional life matches the trajectory of modern burn care. He brought an undergraduate degree from the University of Illinois, an MD from Harvard Medical School, and an eye for the big picture to a long and illustrious career. He was a pioneer in the prophylactic use of antibiotics, which made burn surgery a far safer prospect than it had been for centuries, and a champion of early excision and grafting of burn wounds, the surgical approach that has done so much to reduce mortality from burns. He was instrumental in the establishment of the Boston Shriners Hospital for Children, as well as the Sumner Redstone Burn Center at MGH, and served as chief of staff of both. He capped his career with the invention of Integra, the first artificial skin substitute.

The common thread, and what Burke says is his "abiding interest, is in infection, but not the usual attack on infection, which is how you cure an infection once you've got it. Since I was a surgeon and no surgeon likes an infection, I spent a great deal of my life trying to figure out how not to have infection to begin with." That is either modesty or reductionism considering he was present at the creation of, and contributed to, most of the advances that add up to better survival rates and far better quality of life for victims of severe burns.

The way he tells it, "The driving force behind the development of the bacteria-controlled nursing units had to do with the simple fact that the infection that was killing 60 or 70 or 80 or whatever it is percent of burn patients was generated not by bacteria that the patient brought into the hospital but from cross infection somewhere along the treatment line, maybe before the hospital, during transport, who knows where." It was an article of faith for Burke that preventing infection was far preferable to treating it once it took hold. At the time, the best anyone had come up with was keeping burn patients isolated in separate rooms, limiting the people who came in contact with them, and requiring those few to wear sterile gowns, masks, gloves, and head coverings whenever they entered the patient room. The practice was time-consuming and expensive, and as an infection-control strategy, Burke thought it did not work terribly well. He came up with a self-contained and portable isolation "room" that answered these objections and allowed easier access to the patient, by both the treatment team and family members.

"About that time, right across the river, the Apollo people were getting their act together." Engineers assigned to the NASA Apollo program were working out of a warehouse building belonging to the National Casket Company, and Burke used to run into some of them at the MIT faculty club. Over beers, Burke presented an earthbound challenge: "It's okay to get to the moon and all that stuff, but why don't you guys do something really useful and help us out by keeping patients from getting infected." What he was looking for was a germ-free environment. "Those guys were busy making environments where they could manufacture components for computers or whatever to look at the moon, and if there was any dust around they got into trouble, so they were very interested in the engineering of dust-free environments. Now since bacteria don't float around by their skinny selves—they float around on pieces of dust—if you got rid of all the dust, you got rid of all the bacteria. So we got these guys all fired up about how we ought to do this. The MGH in those days owned a garage on Charles Street and we talked them into allowing us a big piece of space on the fifth floor of the garage and we began to mock up germ-free, bacteria-controlled units."

The end result was a stunningly simple design. The units, which are custom made in the Department of Biomedical Engineering model shop at MGH, consist of a clear plastic curtain wall measuring six by ten feet that completely encloses the patient's bed, separating it from monitors and other equipment, other patients and their visitors, the nurses, doctors, and others who care for them, and the microbes that have taken up permanent residence in every hospital in the world. The temperature and humidity of the air inside the unit is controlled separately from that in the room. The norm is 88 degrees Fahrenheit and 80 percent humidity, but both can be adjusted upward or downward if necessary. Machinery located two floors above Bigelow 13 continuously draws air through a high-efficiency particulate air (HEPA) filter, returning heated, humidified, and bacteria-free air at an exchange rate of 300 times an hour. The mechanics rely on the principle of laminar flow, which delivers clean air and removes used, and potentially contaminated, air in one continuous, steady, nonturbulent, and unidirectional flow.

The climate inside the BCNUs is close to that of an Amazon rain forest, near perfect for a badly burned patient, near torture for the nurses who minister to those patients for hours on end. The full-dress uniform consists of scrubs covered by a yellow plastic pinafore-style apron and wrist-length latex gloves layered over clear plastic gauntlets that reach to the armpits, veterinarian's gloves intended for use in the barnyard but perfectly adapted to the task at hand. Nurses stand outside the plastic curtain wall and reach into the unit through overlapping panels that form an antimicrobial maze. They are able to accomplish an extraordinary range of tasks without entering the unit itself—taking vital signs, suctioning airways, examining wounds, bathing, changing sheets, not to mention doing the dressing changes that can take several hours, and always touching and talking to their patients, maintaining human contact, therapeutic touch. Occasionally they add mask, head covering, and gown to their get-up and go all the way inside if they need to be closer than arm's length away. Working inside the BCNUs, even only up to the shoulders, is like running laps in a steam bath. Bob Droste told me it is not unusual for a puddle of sweat to collect in the elbow bend of his gauntlets in the course of a dressing change.

Still, he said, it is far better than the first week on Bigelow 13. Burke tried out the first BCNUs at the Shriners Hospital, which opened in 1968 with Oliver Cope as chief of staff and Burke as chief of surgery. When Cope retired the following year, Burke took over as chief of staff, a position he held until 1981, when he resigned to become chief of all the trauma services at Mass General. When the burn unit moved from White 12 to Bigelow 13, the single rooms with their banging screen doors were exchanged for four tents in the big ICU in room 1322. Bob Droste was there for the transition.

"We came here in 1984. We brought a couple of sick patients with us. It was kind of a big move, and we had to learn how to take care of them, so we had one week of no plastic. But in the interim, in this room, instead of being relatively comfortable like it is now, the heat was turned up to eighty-plus degrees and the humidity was way up too. To keep the patients warm we had it *real* warm. One day I lost four pounds during a dressing change and four and a half another one. I weighed myself before and after two different days. Four pounds once and four and a half another, in sweat." Finally, Lynn Bellavia said, "We can't do this; it's not possible to take care of patients in these BCNUs with the rest of the room so warm." So after that first sweaty week, the temperature in the room was turned down to bearable levels while the BCNUs were individually adjusted to the needs of each occupant. All it takes is a call to engineering to raise or lower the temperature or humidity within minutes.

The BCNUs are permanent fixtures at MGH and Shriners, and a similar setup has recently been adopted by the Firefighters Burn Treatment Unit at the University of Alberta Hospital in Edmonton, but you will not find them at any other burn center in North America. If they work so well, why doesn't everyone use them? Burke has his salty way of answering that question: "First, it's not an unbelievably expensive operation materials-wise—that is, the walls and whatever wouldn't break many banks, or it would have to be a pretty small piggy bank to be broken by the charge—but the space cost is very high because you have to have not only the ward space but space above to efficiently handle the laminar flow and keep it laminar. And then you've got to have the medical driving force to really believe that this is going to make a difference and that infection isn't inevitable. Finally, you have to have a nursing service that's not only adept, but really highly adept technical people. The nurses here are really great with it. They've been doing it since they were knee-high to a duck and if you tried to stop them, I think you'd have trouble."

CHAPTER SIX

In the Hyperbaric Chamber

The elevator eased to a stop on the eleventh floor of the Massachusetts Eye and Ear Infirmary. A patient transport associate nodded to his fellow passengers, then pushed the gurney through the opening doors. He was surrounded by a phalanx of equipment on wheels, each with its own keeper. A nurse rolled the IV stand and pump, its plastic bags and vials swinging with the forward momentum; the resident piloted the monitor, his eyes trained on the display, his ears alert for any change in the beeping rhythm, the respiratory therapist worked the AMBU-bag while another transport associate brought up the rear with the mechanical ventilator. Like a barge with tugboat escort, the gurney rounded the corner and rolled down the hall toward the Norman Knight Hyperbaric Medicine Center.

The Massachusetts Eye and Ear Infirmary looks out onto the Charles River as it takes the dogleg turn south between the cities of Boston and Cambridge. A tunnel connects it to the Vincent building, making for an easy roll from the MGH Emergency Department. Most people probably do not even realize Mass Eye and Ear is a separate institution, and if they do, they may have only the vaguest idea what goes on inside it. Cataracts? Glaucoma? Hearing aids? Few would imagine that a hyperbaric oxygen treatment facility is tucked into a corner of the eleventh floor or have any idea what oxygen under pressure has to do with eyes and ears.

Hyperbaric oxygen (HBO) chambers are best known for the dramatic rescue of undersea divers suffering the bends, the painful and potentially deadly consequence of a too rapid ascent from the high-pressure depths of the ocean. Although rare today, the bends was a deadly occupational hazard among sandhogs digging tunnels beneath rivers and other bodies of water. So-called caisson disease resulted when the caisson, a watertight compartment that carried the crew to and from the depths, rose to the surface too quickly, usually when there was a collapse or other disaster in the underwater construction site. Also called decompression sickness, the bends develops when nitrogen, which makes up more than three-quarters of the air we breathe, is forced out of solution in the blood. When pressure is decreased gradually, the nitrogen redissolves. When the decrease is sudden, it is like opening a can of soda, except the gas is nitrogen, not carbon dioxide, and the fizz amounts to many tiny gas bubbles inside the blood vessels of a human being. Decompression sickness is hardly an everyday occurrence and is certainly not common enough to support three chambers and the personnel to operate them, twelve hours a day Monday through Friday and, if need be, any hour of the night, any day of the week. Nor does it explain why there are hundreds of such facilities in the United States, many of them far from coastal areas where divers might require them.

Why does Mass Eye and Ear have hyperbaric chambers? The answer depends on whom you ask. Lorraine Brennan, the nursing coordinator for the HBO center, said, "It makes a lot of sense, since we treat everything from the neck up here." That includes head and neck cancers as well as disorders of the cranial and facial nerves, speech and hearing disorders, traumatic injuries, and congenital and developmental abnormalities. "A lot of our patients have conditions that can be helped by hyperbaric oxygen." Ask the same question at MGH and you will hear it is because "they got the grant."

The hyperbaric center at Mass Eye and Ear is a single large room where specially trained nurses watch over the occupants of the clear acrylic cylinders riding on gun-metal blue steel chassis and lined up in a row like something out of *2001: A Space Odyssey*. Despite looking like they are cranked up for takeoff, with their sleek curves and flashing digital controls, the chambers make no noise, not even a telltale hiss or whoosh. Except for the occasional emergency, most patients are conscious, with nothing to do but lie quietly on their backs, covered by a light blanket, dozing or watching the small television hanging above the foot of each chamber, the mundanities of daytime TV flickering in the middle distance. Nurses speak softly among themselves as they check the dials and gauges, and occasionally talk to a patient, using the telephone handset mounted on the side of the chamber. The patient's answer is broadcast through a speaker in the console. The ability to see and to communicate as well as the quiet and the freedom to move—even if just to bend a knee or scratch an itch—keep a session in the chamber from being as claustrophobic as the journey through the clanging confines of an MRI. If the occupants feel anything, it is the slight pressure on the inner ear familiar to airplane travelers, easily overcome with a wiggle of an open jaw or a hearty yawn.

To look at the chambers you would think nothing is going on. Were it not for the digital readout on the control panel, you would never know that the atmospheric pressure inside the chamber is 44.1 pounds per square inch (psi), or 3 atmospheres (atm), three times as great as at sea level. Whatever the therapeutic benefits, and they are the subject of much debate, an enclosure filled with oxygen under pressure is more or less an incendiary device. All it would take is a tiny spark to set off a fire and then, as soon as the oxygen inside the chamber is consumed, an implosion. That is why occupants must adhere to a strict dress code: 100 percent cotton clothing, gowns, bandages; no polyester or plastic anything. No eyeglasses or contact lenses, no dentures or jewelry, except for the antistatic grounded wire bracelet they all must wear. No glass of any kind. No hair spray, perfume, or deodorant, which contain alcohol and other flammable substances; no lipstick, makeup, lotion, or oily hair products, and nothing made with petroleum jelly, which could fuel a fire. Not that such a catastrophe has ever occurred here, but that is because the personnel who run the chambers know what they are doing. They also provide a long list of instructions for every patient who comes for treatment, conscious or otherwise. Among them is the prohibition of carbonated beverages within an hour of treatment, and the requirement that colostomy bags be emptied, should a patient have one. Lorraine Brennan blushingly relied on a euphemism to explain why.

* * *

The quiet calm of the room was punctured when the crew from the General burst through the double doors, bringing the hubbub of the ER along with them. Two of the chambers were occupied, a frail old woman was lying on her back in one of them, a pillow propping up her head so she could see the TV screen. She suffered from osteomyelitis, a painful inflammation of the bone marrow, which was being treated with a combination of antibiotics and hyperbaric oxygen. In the other, a stocky man with diabetic ulcers stretching from his heels to his calves lay snoozing in the silence of his enclosure.

The third chamber, at the far left of the row, had been in use up to a quarter of an hour before, its former occupant now dressed in street clothes and bidding his nurse good-bye. He had only three more appointments before the end of his therapy to rebuild the bone of his jaw, which had been eroded by radiation treatment for cancer. He had been coming twice a week for the past year, his health insurer paying $800 a pop for the treatments. When the call came from MGH that a chamber was needed for a case of carbon monoxide exposure, there was no hurrying him out. It can take as little as seven or eight minutes and as long as an hour to get down to 3 atm. Instinct is to think of it as up, but a session in the chamber is referred to as a "dive" and in terms of the effect on the body, it is the equivalent of descending eleven fathoms below the surface of the ocean. The amount of time it takes to reach bottom depends on patient comfort. If a patient is awake, as this man was, a rapid descent would be extremely uncomfortable and could risk bursting his eardrums. Like most conscious patients, he was given a decongestant to ease the "ear-filling" sensation and taken down slowly, over the course of about an hour. When patients are unconscious, as Dan was, getting them down quickly could be life saving, so their eardrums are commonly pierced beforehand to prevent the buildup of pressure. Coming back up to 1 atm is generally a little faster than the dive, but ninety seconds is the absolute fastest it can be done, and then only in an emergency. An emergency for the person in the chamber, that is.

So Dan had to wait. He did not leave the Emergency Department until 10:00 A.M., but there was plenty to keep everybody busy while waiting for the go-ahead. It took awhile to ensure that he was hemodynamically stable—that his heartbeat and blood pressure were in the normal range, that blood was not only getting to his extremities, but was also coming back. It also took some doing to get him fully sedated. The panic his parents witnessed—recorded as "agitation" in his patient record—had to be tamed with pain and anxiety medication. Patients on mechanical ventilators have to be knocked out to keep them from "fighting the vent," which can range from dysynchronous breathing—exhaling when the machine is delivering air, inhaling when it is pulling air out—to self-extubation. And then there were his dressings. He had to be bandaged in accordance with the requirements of the chamber: nothing flammable, no ointments of any kind. Even soot that might have adhered to his scorched skin had to be thoroughly washed before his wounds were dressed in gauze. It can take an experienced burn nurse well over an hour to wrap a patient with burns the size of Dan's.

The patient originally scheduled for the vacant chamber had been given the choice of what could easily be a four-hour wait or coming back another time. For Dan, every minute counted, so he was moved to the head of the line. The man from patient transport handed over the transfer papers, including the eighteen-item checklist attesting that Dan had been properly prepared for the chamber and all his needs could be provided for while he was there. Then they lifted him and his array of plumbing onto the special HBO-worthy bed, rolled it in through the open end of the chamber, and positioned the tubes and hoses in their assigned ports. Finally the hatch was closed and tightened like the airlock on a submarine. Dan was visible through the clear plastic, lying as still as Tutankhamun in a space-age sarcophagus. The respiratory therapist would stay in the room the entire time, monitoring his breathing and making adjustments if needed.

Dan was programmed for a ninety-minute dive, the standard for each session, but he would be in the chamber for at least two hours, between ten minutes down, maybe half that coming up, and two obligatory air breaks of ten minutes each. As much as a person in respiratory distress needs oxygen, too much can be as bad as too little. Under normal pressure, breathing 100 percent oxygen does not usually cause problems unless it is for longer than twenty-four hours. Under hyperbaric pressure, however, nausea, dizziness, blurred vision, and involuntary muscle twitching may develop in an hour or less, and there is the potential risk of seizures, collapsed alveoli, and even blindness. Oxygen toxicity can be avoided with periodic whiffs of room air—21 percent oxygen, 78 percent nitrogen, the remaining 1 percent being a mix of gases running from argon to xenon. A nurse picks up the phone to tell patients who are conscious and breathing on their own to put on a face mask while she changes the gas input to room air. For patients on mechanical ventilation, the respiratory therapist adjusts the ventilator settings to a less oxygen-rich mixture. The pressure, however, remains at 3 atm.

* * *

Aside from treating the bends, HBO is used for a range of medical problems caused or fostered by a shortage of oxygen. One example is life-threatening bacterial infections, such as those caused by rampant *Clostridium perfringens*, an anaerobic bacterium that thrives in the airless atmosphere of a festering wound. Slamming the wound with a souped-up dose of oxygen prevents *C. perfringens* from producing the toxins that cause gas gangrene, something a boatload of antibiotics cannot do without risking side effects difficult for a very ill person to withstand. Lower doses risk the development of antibiotic resistance. Combining HBO with lower doses of antibiotics is seen as a safer strategy. Other applications include wound healing and tissue regeneration, which require plenty of oxygen in an area that tends to be oxygen poor. It is not the skin that gets the benefit of the extra oxygen; it is the blood. Hemoglobin takes up oxygen more readily and in greater volume when it is delivered under pressure, producing a superoxygenated fluid. Many of the people who come to the HBO center at Mass Eye and Ear suffer from chronic wounds that heal slowly or not at all. This is commonly due to circulatory problems related to diabetes. Others suffer from osteoradionecrosis, degeneration of the bone after radiation treatment for cancer. Oxygen-enriched blood is thought to help revitalize the blasted area.

It sounds as though burn wounds could benefit from the same sort of treatment, and indeed those who promote HBO therapy tout it as a healing accelerator for both acute burns and skin grafts. Within the burn care community, however, this claim is regarded as questionable. HBO is rarely, if ever, used for burn wound healing. The only purpose for which it is even occasionally tried in fire victims is to combat the effects of carbon monoxide exposure.

Carbon monoxide (CO) is colorless, odorless, and tasteless, making it undetectable by anyone breathing it. It interferes with oxygen delivery to tissues in two ways. First, it hogs the bonding sites on hemoglobin, the vehicle that carries oxygen in the blood. Blood continues to circulate, but much less oxygen (O_2) is available for use by the body. Second, it obstructs enzymes that enable tissues to use what oxygen they do get. The resulting hypoxia is particularly damaging to the brain, liver, and kidneys. When carbon monoxide binds to hemoglobin, it forms a compound called carboxyhemoglobin, or COHb. The COHb level is expressed as a percentage of total hemoglobin. For example, the COHb level in a non-smoker breathing noncontaminated air is under 2 percent. One- to two-pack-a-day smokers usually have COHb levels of 4–5 percent. Upward of 20 percent is toxic and anything above 50 percent is potentially lethal.

The COHb saturation depends on how much carbon monoxide there was in the air and how long the victim was breathing it. It can take as little as three minutes in a densely smoke-filled area to build up a COHb level of 30 percent. The Melody Lounge in the basement of the Co-coanut Grove is a perfect example of such an environment; so is the small bed-sitting room where Dan was found by the Boston fire department. Symptoms begin with a slight headache and impaired judgment, and intensify to nausea, vomiting, and dimming vision as COHb levels increase. Loss of consciousness and coma usually occur at 50 percent; death arrives in a matter of hours at 70 percent and a matter of minutes at 90 percent. When it does not kill, carbon monoxide can cause significant brain and nerve damage, some of it temporary or reversible, much of it permanent. These may include changes in behavior, memory, and cognition, as well as fine and gross motor disorders. Sometimes the damage is apparent early on, though it might be difficult to determine that with a patient who is unconscious and immobilized. Sometimes no brain damage is apparent at first, but signs of dementia emerge one to three weeks after a "lucid interval," during which the patient appears to have recovered totally from CO poisoning and shows no neurological effects.

When Dan O'Shea's COHb came back at 46.2 percent, John Schulz had reason to be alarmed. Theoretically, as soon Dan was dragged from the fire, even before the paramedic slapped the oxygen mask on his face, his body began to clear itself of carbon monoxide, though the fact that he was barely breathing slowed the process. The half-life of carbon monoxide in fresh or uncontaminated room air is about 250 minutes. That means a person who has been exposed to carbon monoxide will be able to clear half of it in a little more than four hours by breathing room air alone, and then half again in another four hours, and so on until it is down to normal. At very low levels of exposure, that might be okay. Upping the oxygen dose by administering 100 percent oxygen reduces the half-life to between forty-five and sixty minutes. Adding pressure to that further reduces the half-life; 100 percent oxygen delivered at 3 atms will clear half the CO in less than half an hour.

Sounds like a very good thing, but there is a catch.

Despite the proximity by geography and affiliation between the General and Mass Eye and Ear, it is rare to see a burn patient in the HBO center. For most burn patients, the risks simply outweigh the benefits, which have not been established to anyone's full satisfaction. The problem is that while a patient is in the hyperbaric chamber the burn team has to suspend everything else it needs to be doing in the early hours after a burn. The American Burn Association Practice Guidelines go no further than to say HBO "may be appropriate" only if the treatment does not in any way interfere with general burn management and only for patients whose COHb is greater than 25 percent, with depressed mental status, who are otherwise stable and do not require ongoing fluid resuscitation.

Many burn centers frown on hyperbaric treatment, contending that six to twelve hours of 100 percent oxygen at normal pressure, which can be delivered in conjunction with acute burn care and with the assist of a mechanical ventilator for patients unable to breathe on their own, is equally effective. Cynics say that centers that have easy access to hyperbaric chambers or have their own are more inclined to regard the treatment as beneficial: If they have the technology, they will use it.

With Dan O'Shea, said John Schulz, there was no debate. "It was unambiguous. He was found down with agonal respirations and his carboxyhemoglobin level was very high. . . . All bad." The issue facing Schulz was not that Dan might die from lack of oxygen. "We can rapidly compete that carboxyhemoglobin away with 100 percent oxygen at atmospheric pressure of 1," he explained. Getting pure oxygen into the lungs even without high pressure is sufficient to prevent asphyxia. "What we're worried about is neuropsychiatric problems afterwards. The 25 percent carboxyhemoglobin cutoff is based on clinical data saying that people *might* do better neuropsychiatrically using hyperbaric oxygen. It's enough of a 'might' that we use it."

Dan had been on 100 percent oxygen since he was pulled from the fire. His blood went to the lab no more than a half hour later. If it was 46.2 at that point, working backward, with a half-life of forty to sixty minutes on 100 percent oxygen, it might have been as high as 75 or 80, enough to kill him outright or leave him with severe and permanent neurological damage. Dan's parents recall someone comparing Dan's exposure to a sixty-five-year-old man who had smoked ten packs of cigarettes a day for his entire adult life, but Schulz told me it was far worse than that: "He was like somebody trapped in a closed garage with the engine running." On the other hand, by the time the results came back from the lab, Schulz could see that Dan's heart was pumping blood to all parts of his body, his resuscitation was progressing, there was no risk of shock, his airway was open, the respirator was doing its job, and he did not need immediate surgery. "Most people are not candidates. They're just too unstable. Somebody's got to be very stable before he's put in there. Fortunately, Dan was, so we decided that hyperbaric oxygen might help him."

They ended up using it twice for Dan, who went back for a second session the next day. "Once is good, twice is better," Schulz said. He was following the lead of Rob Sheridan, who holds an appointment as attending hyperbaric physician at Mass Eye and Ear. Among the three Burn Associates, he has the most experience with HBO treatment, according to Schulz, and it had been his practice to repeat treatments. "Nobody really knows. We have treated somebody three times. We treated Dan once, he tolerated it fine, so we did it again."

Still, the chamber makes Schulz nervous, if for no other reason than the impossibility of a quick evacuation for a patient in distress. An HBO chamber is kind of like a giant pressure cooker. You cannot just open it up before you bring the pressure back to normal, and you cannot do that in a hurry. The folks at the HBO center say they can get up to 1 atm very quickly in case of emergency. Schulz is not persuaded. "Right. You can get them out fast, but they might get the bends."

Bob Droste is no fan of HBO either. Burn nurses do not get to decide whether a patient goes in, but they are the ones who take care of the patient before and after the treatment and, if need be, during. In addition to the respiratory therapist, a nurse from Bigelow 13 has to stay with patients in critical condition the entire time, putting a squeeze on the rest of the nursing staff. Being there, however, does not mean there is much a nurse can do. The chambers cannot be heated so it is hard to keep the patient warm. Wound care is obviously out of the question, as is any of the other hands-on duties nurses perform. So it is mostly a matter of sitting there, hoping for the best and dreading the worst. "We did once have a patient arrest in it, and the nursing staff almost lost their minds waiting for them to get her out," Bob said. Fortunately—surprisingly—the patient ended up being all right. Her arterial blood was so totally saturated with oxygen that she did not succumb to two minutes without a heartbeat.

"Dan came in at a time when we were a little bit more optimistic about using the therapy," Schulz acknowledged. "I'm really reluctant to put a badly burned person in there now, because you just can't get them out fast enough. But it's rare for us to see somebody whose carboxyhemoglobin is as high as his was. The higher it is, the more incentive to get rid of the carbon monoxide as fast as you can. It definitely has uses, but it's a dangerous place for a sick person to be, in my opinion."

* * *

It was after 2:00 P.M. when Dan came up from his dive, but there was still no bed for him on Bigelow 13, so back he went to the ER, patient transport leading the convoy, the fleet of equipment following in its wake.

John Schulz was upstairs trying to figure out which one of the critical patients could be moved out of the unit to make a BCNU available for Dan. "There was a real bed crunch when he came in, so we were trying to triage people around," he recalls. Sometimes the triage is easier than others. Sometimes a patient is nearly ready to move out onto the floor and this will become the occasion; at other times, no one is a good candidate and it may involve sending someone to an ICU on another floor.

Schulz remembers the time Dan came in as a particularly difficult one for the staff. "We had a lot of young, badly injured people." He said that is the reason it sticks in his mind so many years later—the harder things are, the more Schulz remembers. And young people with bad burns are the hardest of all.

The ones it was hardest on were Jack and Peggy O'Shea. They sat alone in the family waiting area through the long morning and longer afternoon. It is not much more than two loveseats and a half dozen armchairs, upholstered in dark blue vinyl printed to look like brocade, a few coffee tables and some lamps, and the standard-issue potted ficus that, on closer inspection, turns out to be plastic. It sits out in the open like the set of a daytime soap on a corridor that bridges the Bigelow and Blake buildings, with window walls on both sides giving a panoramic view of the city. It also provides a ringside seat to a parade en route to and from heaven knows where. People came and went, but the O'Sheas kept to themselves, unable to speak, even to each other, or to think about anything except what had happened, what they had seen, what would come next. Sometimes someone would park an empty gurney against a wall. Or the clunk of the clogs everyone in the hospital seemed to wear would herald the arrival of one or two figures in scrubs who might nod hello as they passed. One time a gurney came by in a rush, four or five people walking rapidly alongside holding bags of clear and cloudy liquid high above their heads. Peggy caught a glimpse of a body lying on the bed. Could that be Dan? What was taking so long? Would this day never end?

Jack paced and sat, paced and sat, while Peggy gazed through the window wall in front of her, talking to God while she watched the winter afternoon sky darken and the lights come on downtown: the Hancock Tower, the Pru, the ribbon of lights along Storrow Drive, and flickers from the boathouses down at the river's edge. It was well into the evening before someone came out to tell them that Dan was almost settled in and they could soon come back to the unit to see him. Neither Peggy nor Jack remembers who it was. It is another one of those details blurred by time and the anguish of the moment. The people they remember, aside from Dr. Schulz, are two nurses who took care of Dan during his long stay, Mary Williams and Mike Wilson.

Mike Wilson is a "lifer." A man in his fifties, he has worked on Bigelow 13 since 1988. He makes no bones about how he spent his life before that. "My first career was as a bum. I was actively drinking and drugging when I was about fourteen. And I was in that world till I was about thirty-four, thirty-five. I was a drug addict, alcoholic, lived on the street. A bum." He says it loud. "Everybody needs a bum."

He is a world-class talker, admits it himself, but he is as articulate as he is discursive, and he grabs you with his fervor. It is the fervor of the convert, but somehow you grant him that, even honor it, because of what he has gone through to get to where he is today and mostly because of what he does now that he is there. In addition to being a highly experienced burn nurse, he is a trained pastoral counselor, one of two on the floor. Along with Frank Ireland, he devoted 400 hours to a clinical pastoral education program, originally designed for clergy but open to laypeople and incorporated into the multidisciplinary team. A study Mike and Colleen Ryan did found it "a helpful adjunct to the traditional physical, psychological and social care of our burn patients." Despite the availability of hospital chaplains, Mike thinks nurses are in a better position to attend to the spiritual needs of patients and their families. He regards that as the most important part of nursing, not "hiding behind the numbers: the blood pressure, the pulse, the medication schedule." Mike's great gift is the way he connects with people—families no less than patients—where they are most vulnerable. His gentle sensitivity to their spiritual state of being involves a level of caring that goes beyond what is usually considered the nurse's mandate. But the best nurses give something in their touch, their voice, their attention, their presence—caring taken to the highest level.

If Bob Droste is the cynic of Bigelow 13, Mike Wilson is the philosopher. Their personalities could not be more different, but they make a good tag team. You get top-notch nursing with both. Bob saves the view from his jaundiced eye for his colleagues. Mike waxes metaphysical at the least provocation to anyone within earshot.

"I believe this is one of the most miraculous places you'll ever want to be in. It alters everybody's life that walks in and out of here. I know how it's altered my life in so many different ways. Why do I think that is? I think that there are gifts that are given to us in pain and suffering, not just here with the burn patients, but anywhere, and the only time most of us ever stop to look or to listen or to think is when we're in pain and we're suffering. And there's so much of it here that it really puts people in touch with that aspect of themselves, and lives are changed tremendously. For the better. I've watched it with patients, I've watched it with families, I've watched it with visitors. I think it's the same kind of thing that's present in the death process, the same kind of thing that's present in the birth process, and I think we're so busy with our lives that we're out of touch with that a lot, but in a place like this, it's in your face. It helps to get in touch with that. You know, burn patients are the sickest of the sick, and you can't help but be put in touch with your own mortality as you do this."

That kind of intensity might not be everybody's taste, but for the O'Sheas, Mike was a godsend, not just because of how well he cared for Dan physically, though he did that too, but because of the emotional and spiritual support he gave them all. No matter what was bothering them, they felt they could talk to Mike about it and he would do what he could to change it or, if he could not, to let them know he shared their pain.

Mike's spirituality was what made a difference for Peggy. "It was invaluable having Mike Wilson there, for me, anyway. He was a touchstone in a way. I do think it is important to have someone who has some kind of a spiritual outlook, so you can touch base with them." Later, when Dan was awake, Jack relied on Mike to talk with Dan about his drinking, to "set him straight," as Jack put it, something he did not feel able to do himself. Perhaps because Jack had experienced his own battle with the bottle, he was not comfortable taking to his son to task about it, but Mike figured his own alcoholic past gave him the needed authority. "I suppose I was looking to him to be preaching to Dan, that Dan would listen to him, you know. I'd say, 'Mike, make sure you talk to him,' and he'd say, 'Don't worry, I've got his number.'" And then he'd go in and say, 'Dan, you know, I'm a drinker myself . . . '"

* * *

They needed all the support Mike and anyone else could give them that first night as they approached the plastic tent, holding their breath as tightly as they held each other's hand. Dan lay inside, wrapped in yards and yards of bandages, his arms stretched straight out from his body. The bed was tilted to elevate his head and chest. Someone explained it was so fluid would not build up in his lungs. There were countless tubes and wires running from his nose, his mouth, from folds in his bandages, out through the bottom of the plastic curtain. One tube was connected to a bag hanging over the side of the bed into which was dripping what could only be urine, two or three were stretched up to a metal stand from which hung a jumble of plastic sacks, a larger accordion hose fed out of a mechanical respirator, chuffing rhythmically as the occasional drop of something—Peggy did not even want to know what it was—dripped into a puddle that was sloshing around at the low point of the hose. Something was beeping—*deet-deet, deedle-deet.* No one else in the room seemed to notice, but to Jack it was an infernal sound and to Peggy a source of anxiety. As was just about everything else.

There were four beds in the room, four tents, and inside each was someone who was really sick, you could tell by the smell, by the buzz of activity
around each one, by the competing beeps and blips coming from the monitors. Could Dan catch an infection from one of the other patients? All that
urine, all those smelly bandages. Was it really safe to have them all in the
same room like this? Across the room was another man. He looked young,
like Dan, and he looked awful. A nurse was unwrapping his bandages,
which were stained brown like dried blood and soggy and sour smelling.
The nurse kept tossing the bandages into what looked like a red laundry
hamper at the foot of the bed, just inside the plastic tent. Sometimes they
went right in, sometimes they hung over the edge and touched the floor.
His skin was black in places, really black, like it had been dipped in ink.
Could all that be burns? Is that what Dan looked like beneath his bandages?

Later Peggy would learn that the stains were silver nitrate, which turns
healthy skin inky black and bandages brown but soothes open wounds
and, more important, guards them from infection. She would learn and almost believe that the plastic walls that separated her from her son were
keeping him safe from whatever it was the other patients might have. Later,
Mary Williams would show her and Jack how to gown up—donning yellow cotton smocks and covering their hands with latex gloves, their
mouths with paper masks, their hair with what looked like paper shower
caps—and then she would pull back the lower plastic curtain so they could
put their hands inside and touch their son, stroke his cheek or bandaged
arm, and speak softly to him in the hope that maybe he could hear.

But for now, Dan lay motionless on the other side of the plastic barrier.
If not for the rise and fall of his chest, you would not know he was alive.
And even that—Peggy knew it was the respirator that was making his chest
move, pushing air into his damaged lungs, then pulling it out again. Her
worst fear was that he would never again be able to breathe on his own.

CHAPTER SEVEN

At the Bedside

"The trained nurse has become one of the great blessings of humanity, taking a place beside the physician and the priest, and not inferior to either in her mission." So declared Sir William Osler, the great physician and educator. But what exactly does a nurse do? All you have to do is watch the morning routine in room 1322 and you will know the answer: Everything.

The BCNUs are designated A through D, clockwise from the first bed in and to the left as you face the back of the room. On this particular morning the unit is filled with a collection of stories that run from sad to tragic—the only kind of stories you hear in this part of Bigelow 13.

Mary Williams is working at A bed, preparing to change her patient's dressings, a twice-a-day ritual that is central to a burn nurse's duties. The hands-on care and the close observation entailed are among the reasons why bedside nurses know their patients better than anyone else on the burn team. From takedown to final wrap, a dressing change may take more than an hour. Two hours is not unusual for a big burn. During that time, the nurse examines every inch of the patient's body, noting any change in the wounds, for better or worse, rubbing, scraping, pulling off dead skin as part of the continuing debridement project, checking for bed sores and oozing wounds, for signs of infection and signs of healing. It is an awesome thing to watch—a miles-long obstacle course run in tropical heat, the nurse a warrior-athlete cradling the patient's inert body from starting block to finish line.

This particular patient's gender is indeterminate, so distorted is the body by the effects of the burn and its aftermath. The trunk is still bloated by edema, the face still swollen, the scalp a mosaic of wounds and patches of hair. The legs seem like sticks that could never support the elephantine torso. In fact, the patient is female, not long out of her twenties. By the look of the photographs tacked to a pillar near the foot of the bed, she was once a smiling, attractive woman, slender enough for a mother of five children perhaps too closely spaced from preschool to preteen. Four of the children are praying for her now. They buried the fifth, the youngest, the day after the car burst into flames, rear-ended at a stoplight by a drunken driver. Her husband, who was holding the baby in his lap, died of his burns two days later.

This family, like so many others who come to Bigelow 13, is in deep crisis, ripped from everyday life by a senseless tragedy. When this patient pulls through—if this patient pulls through—she will face months of rehabilitation before she can return to a semblance of normal life, and she will return to a family deeply scarred emotionally by its losses. The sadness of the situation is unspeakable, and it touches everyone who comes in contact with the patient and her family.

Mary Williams sees her work with the family as integral to patient care. "In order to really give good care, you have to get to know the family. Because the patient doesn't know much what's going on. They have physical pain, and you can take care of that. But the mental pain, that is what the family has. Imagine that you are going to leave your loved one here and he may not make it through the night. You need to develop a relationship with someone who you know is giving the best. It's only by developing that relationship that you can go home and rest comfortably knowing that your relative is in good hands."

Mary spends a lot of time with this woman's family—her mother and her sister, who are taking care of the children—when they come to visit, and she speaks with them when they phone in the morning. It is the same with all the nurses. More than one patient who has progressed from BCNU to floor, from drugged comalike sleep to increasing awareness, has expressed surprise at how well the nurses know their family members and how strong a bond has formed during the dark days they know nothing about. Nancy Parent knows about that bond, even though Tom was awake the whole time. One of the nurses told her, "Usually, we're more involved with the patient's family than the patient because most of our patients are out of it. So by the time they wake up, they don't understand how we've become so close to the family, but it's because we're working with them all the time."

A tall woman with a regal bearing and a trace of the Islands in her voice, Mary is older than many of the nurses on Bigelow 13. She is a grandmother with grown children, one of them a doctor, but at the bedside she looks as leggy and agile as a point guard for the WNBA, and it is clear how physical the work is, how much strength and stamina it takes.

The first thing she does is check the microinfusion pump to make sure there is enough pain and antianxiety medication to get through the procedure. Dressing changes can be painful, even for patients in the twilight world this woman inhabits. Mary punches in some numbers on the touch pad and the fast-acting Versed joins the morphine in a precise, continuous drip. The pump will sound an alarm when only a few minutes' supply remains. She checks the thermostat to make sure the temperature inside the BCNU is on the high side since she is about to literally undress her patient and expose her evaporative surface to the air. High humidity and high temperature will minimize the heat loss.

Next she suits up for the task ahead. She plucks a visor from the collection hanging from the swing arm of the overhead monitor and pulls it down and back over her brow, ties a fresh yellow plastic pinafore around her waist, dips her arms into the barnyard gauntlets, and snaps a pair of latex gloves over them. Finally, she pulls back the elastic top of the bottom sidewall curtain of the BCNU and reaches into the steamy interior.

An array of tubes and bottles sits on a movable shelf attached to the overhead bed frame. Mary pushes the shelf back toward the foot of the bed and begins to load it with autoclaved bundles of supplies wrapped in blue paper toweling: a stainless steel bowl filled with sterile scissors, Kelly clamps, forceps, and assorted other instruments; a wash basin; two batches of gauze sponges; innumerable rolls of Kerlix, the gauze wrapping Bigelow 13 goes through at the rate of close to 40,000 yards a year; a roll or two of stretchy fishnet that is used to hold dressings in place; a supply of specialized bandages—a full suit of absorptive dressings, huge gauze sandwiches consisting of a front and back vest, and one shaped for each arm and leg, which spare nurses precious time when there are large areas to cover; individually wrapped silicone, hydrogel, and foam pads that protect delicate tissues on the road to healing at the same time as they keep the wound from drying out; a tube of enzymatic debriding ointment. She pulls out of the tent, beads of sweat already forming on her upper lip, and reaches for two bottles of sterile water and one of silver nitrate solution, which she has taken from the warming cabinet.

Incongruously, Janis Joplin's voice wafts out of the stereo speakers on the windowsill, wailing "Piece of My Heart." The unit works to the tune of whatever strikes the fancy of the person who gets to the CD player first. Mike Wilson goes for classical guitar, and when the situation warrants, he puts on what his colleagues call "Mike's music to die by," but classic rock is the usual fare. One hapless junior resident once queued up "Dancing Queen" and was roundly jeered by the assembled company, which declared the unit an Abba-free zone.

Mary begins unwrapping yards and yards of Kerlix from her patient's limbs. As she gets to the end of each damp, grayish length, she tosses it into the red hamper at the foot of the bed. Slowly she reveals the arms, the hands, and then moves down to the legs. A patchwork pattern of meshed skin grafts covers most of the woman's legs, tailing off as it reaches the thighs. The woman's feet are cradled in splints, fitted by the occupational therapist, whose expertise includes positioning burned limbs for maximum healing and minimum loss of function. Many splints are custom-made, tailored to a patient's particular need, but these are the standard-issue L-shaped type, lined with egg-carton foam and held in place with Velcro straps. Mary removes the splints and puts them up on the caddy, then unwraps each foot in turn, revealing blackened toenails, and on the heel of one foot, an oozing wound. Black toenails are normal, discolored by the silver nitrate solution that drenches the gauze and wards off infection at the same time as it turns everything it touches a minstrel-show hue. The oozy heel is not. It is what nurses call a pressure ulcer and the rest of us know as a bedsore. Mary makes a mental note to give it further attention.

She raises her patient's left leg, slipping it into one of the Ace bandage loops that hang from the overhead frame of the bed, and begins to gently wash it. Reaching over to the caddy, she empties a bottle of warmed sterile water into the stainless steel basin and moistens a gauze sponge, then wipes it over a small section of skin before tossing it into the red hamper at the end of the bed. Over and over, she dips, wipes, slam dunks, sometimes rubbing in a circular motion to loosen a bit of dead skin, pulling with her gloved fingertips or forceps, clipping with a pair of scissors from her bowl of sterile instruments. The nurses call a patient like this a "picker's dream." The picking is not painful. It is like pulling off bits of peeling skin following a sunburn. Every once in a while she feels a surgical staple that escaped removal. Skin grafts are held in place with dozens and dozens of stainless steel staples. Over the course of time, the staples are removed, but it may take more than one pass to find them all. As many as possible are taken out when the grafted area emerges from an undisturbed week under a thick quilted dressing, but there are always a few that go unnoticed until a nurse feels them during a dressing change.

Mary reaches for a staple remover on the shelf above the bed and deftly picks out one, then another, then passes her index finger over the area, feeling for other fugitive bits of wire. She takes a look at the ulcerated heel, dabbing it dry, then covers it with a gel pad to cushion it against friction from the splint and give it a chance to heal. Later she will cover the raw areas of scalp in the same way, and when she is done, position the woman's head on a Gel-E donut, a small, gel-filled pillow that guards against the breakdown of fragile skin stretched over bone that is almost inevitable when patients lie motionless for weeks on end.

A circle of sweat has formed on the back of Mary's uniform as she continues her survey and scrub—the other leg, the arms from axillae to fingertips. As she finishes each area, she covers it with toweling to keep the patient warm.

Carefully holding the tubes and monitor leads out of the way and deftly balancing the accordion hose of the respirator, she rolls her patient onto her side, unties the back piece of the gauzy vest, and tosses it into the red hamper. A perfect pale pink rectangle covers the back from shoulder to waist where healthy skin was harvested to provide the grafts for the legs. This is the donor site, and it is healing nicely, though the new epidermis is still delicate and would be painful to the touch were the woman not well medicated. Mary swabs the area gently, then positions a new vest back and rolls her patient onto it.

Nearly an hour has elapsed and the routine is about half over, though doing dressings is but a fraction of what a nurse must accomplish during a single shift. When schedules and staffing allow, expert aides known as burn techs assist or take over the dressing change entirely, but this morning Mary is on her own while the two techs working this shift are tending patients out on the floor.

Across the way in B bed, Sally Morton has just finished bathing her patient and changing his bedding, a complicated business given that he is unconscious, immobile, and attached to a respirator. He has a feeding tube in his nose, a breathing tube in his mouth, a central line below his collarbone and a peripheral line running into his right arm, an indwelling catheter and temperature probe in his nether regions, EKG probes clipped to his skin beneath the dressings that cover his chest, and a cervical collar to stabilize his neck, which was injured when he jumped from an upper story window of his blazing rooming house three weeks earlier.

A yellow cloth hamper stands next to the red one at the end of the bed, overflowing with the wine-red sheets that cover the inflatable mattresses throughout the floor. Red is for medical waste and is clearly marked "biohazard"; yellow is for linens that will be laundered and returned to service. "I think if I went to hell and they gave me a job, it would be to work in the linen room," Sally says with a laugh.

Sally is the only nurse in the unit who regularly wears a dress rather than the two-piece pajama scrubs. Even though it is hospital blue, her outfit is vaguely reminiscent of the crisp white uniforms nurses used to wear. Even her stockings are whitish, but her clogs and the blue paper pouf that covers her dark, shoulder-length hair dispel the illusion that she has dropped in from an earlier era.

She is preparing to bring her patient down for an MRI. The doctors are concerned about possible brain damage resulting from his fall. X-ray technicians come up to Bigelow 13 every day and sometimes more than once. Pushing a huge portable machine, they do the rounds of the BCNUs, typically during the overnight shift so films will be available for morning rounds. But when a patient needs an MRI or CT scan, it involves traveling off unit, a major undertaking.

Sally stands at the cabinet shared by the two nurses on this side of the room. It holds basic supplies and equipment, and a separate red biohazard receptacle for used needles and blades. She is tossing tubes and vials and syringes into a clear plastic shopping bag, the sort used to hold patient belongings. "Today's a travel day," she announces. It looks for all the world like she is packing a toilet kit for an overnight trip. It will not be overnight, but it could take half the day. During that time Sally's patient will need IV fluid, liquid feed, the drugs that are keeping him out of pain and out of touch. He may be gone long enough to need his twelve o'clock meds: vancomycin, the WMD-strength antibiotic prescribed for his recurrent pneumonia, and Reglan, which keeps his gut moving. She has already given him potassium to adjust his electrolyte balance, and that ought to hold him until they get back. She tosses in a few extra syringes, a catheter line, a hand-suction bulb, ticking off the list in her head.

She also has to be prepared for the worst—cardiac or respiratory arrest. Into the bag go the code drugs, atropine and epinephrine. Patients travel with at least two people—a nurse and someone from patient transport is the skeleton crew—so if an emergency arises, the nurse will start resuscitation while the other person rushes to the nearest house phone to call a code blue, the signal that a patient is in mortal danger. Until the code team and crash cart arrive, however, the nurse is "it" and every second counts.

While she is waiting for transport to arrive, Sally notes her patient's temperature—a little low at 98.1—so she covers him with a quilted foil space blanket. Because he will be away from the special environment of the BCNU for a good chunk of the day, she puts an electric warmer on the bed that can be plugged in once they arrive at the MRI suite. Sally will stay with her patient the entire time, and it may be hours before they return to the unit. Meanwhile, that is one fewer nurse on Bigelow 13 at a time when the staff is already stretched thin.

* * *

Tom Parent still remembers his first morning on Bigelow 13. "I was having a really hard time. Not that I couldn't eat, but everything was bandaged and I couldn't work anything." That was when Nancy came in, gowned and gloved, a puffy blue paper cap covering her blond hair, a mask over her mouth. Earlier, the phone had rung in her hotel room. It was the nurse assigned to Tom. "She said to me, 'He's awake. I'm short staffed today, so I wonder if you would mind coming in, feeding him, and helping him clean up a little bit.' I'm sure it was a complete lie, but it doesn't matter. They're very smart. They understand what you're about to go through before you do. They put me to work, and that was the best thing they could have done. So I wasn't in there helpless and blaming myself."

Tom laughed. "They were never really short staffed, were they?"

"No, I don't think they ever were."

Little did Tom and Nancy know.

Except for the doctors, everyone on Bigelow 13 works with both burn and plastic surgery patients. That includes the support staff, nutritionist, occupational and physical therapists, speech-and-swallow specialist, and especially the nurses. It means nurses have to maintain their skills at the very highest level in both subspecialties, but that is something nurse manager Tony DiGiovine insists upon so he can move staff around depending on what the patient census and needs are at any given time.

Tony is a man with a lot on his plate. Although he is hardly the only person on Bigelow 13 who qualifies as overworked, it is still astonishing that he manages not only both sides of the floor but also the transplant unit on the sixth floor of the Blake building. The secret may be that he is not a micromanager. "I *can't* micromanage," he insists. "I don't have the time." Instead, he relies heavily on two women to keep it all running. Carolyn Washington is the operations coordinator for both Bigelow 13 and Blake 6, managing a staff of thirty-two and the supply and materials budgets for both units. She is arguably the second busiest person on a floor where no one is a slacker. "I'm Tony's partner. I follow him everywhere he goes," Carolyn said.

She has been on the scene since burns were over on White 12, having started at the General in 1966, working weekends to earn enough to buy a washer-dryer from Sears Roebuck. She tried out a number of departments—dietary, X-ray—and rose to the rank of assistant supervisor before leaving for Detroit in the early 1970s. She used to come back to work in the summer, but when she came back for good in 1979, the only full-time opening was a secretarial position on White 12. She thought it was awful at first, but soon got used to the rigors of working in that environment. Like everyone else who was there at the time, the first thing she mentions is the screen doors. The second is Dr. Burke, whom she refers to as "my best friend." She became unit coordinator in 1981 and was instrumental in the move from White to Bigelow in 1984.

Washington is the only African American in a managerial position on Bigelow 13, and one of only a handful in any capacity. Her staff—the unit service associates and operations associates, the General's euphemistic titles for the housekeepers, receptionists, record clerks—are all black. The

General has a problem that reflects Boston as a whole, one of the most racially divided cities in the North with a painful history of racial strife in the twentieth century that stands in stark contrast to its years as the vital center of the abolitionist movement in the nineteenth. The hospital deserves praise for its institutional efforts to enhance diversity, which include a multicultural affairs office with a mission "to facilitate and promote the advancement of URM [underrepresented minority] physicians and aspiring physicians and researchers, as well as to develop culturally competent physicians, at MGH," and similar initiatives for nurses and other patient care professionals. Nonetheless, minorities are overrepresented among porters, food workers, custodians, clerical workers, and other lower-level jobs, and underrepresented in the professional and managerial ranks. On Bigelow 13, the occasional nonwhite resident passes through—Jennifer Verbesey's senior, Dave Cooke, was one of the few—but among the permanent patient care staff only Mary Williams and Edna Gavin, a burn tech, are African American. Like Carolyn Washington, both have had long tenures there and are held in the highest esteem by their colleagues, but the disparity is striking all the same. As for other racial or ethnic minorities, they are as good as invisible.

Tony DiGiovine is painfully aware of the problem, but his first priority, in the face of a severe nursing shortage nationwide, is to hire candidates, regardless of race or ethnicity, with the education and, he hopes, the fortitude to work as burn nurses.

"I probably have a 20 percent vacancy right now," he told me in 2002. "Historically we've had a very low turnover rate, but in the last fifteen, sixteen months, we've had a significant turnover, and most of that has been people who've been here two years or less."

When people leave, Tony said, "There are a number of reasons, and not all of it is because of job dissatisfaction, or because they're tired or burnt out. People leave for salary differences. We compete with our own institution in other areas where they don't have to work off-shift or weekends. People will go to work for an agency, because the agency pays at a higher rate and they supplement their housing. People leave the state because significant others leave, cut hours back because they have children and they want to work part-time. This profession allows that kind of mobility and flexibility, but many positions have gone unfilled because we just haven't seen candidates. The class just graduating is where I've seen the bulk of the candidates, so there is not a lot of experienced staff coming to work here."

The nursing shortage is a headline maker and is viewed as a looming crisis, the next hit our teetering health care system will have to withstand. The American Hospital Association estimates that there are 126,000 unfilled nursing positions nationwide and predicts that figure will double over the next decade. Yet fewer young people are choosing nursing as a profession. Nursing school enrollment was about the same in 1995 as it was in 1985, despite the burgeoning patient population, and it is down 20 percent since 1996. That mirrors Tony's vacancy rate and casts doubt on the chances he will fill those slots anytime soon. Hospitals in Boston to Bellingham and every town and city in between are desperate for nurses and recruiting like crazy. The prospect of working at a hospital with the reputation of the General might be attractive; working in burns is not. Over and over you hear the same thing: "These wounds are pretty tough to look at and some people just can't stomach it." "It takes a special person to work in a burn unit." "If you don't like wounds, you've come to the wrong place."

It is not in Tony's interest to make it seem any easier than it is. "When I hire staff, I tell them, 'This is an extremely physically and emotionally demanding area to work in.' There are easier places to work. I try not to scare people away, but many people have no clue. I've had people who think they want to be here, and near the end of their orientation, they come to me in tears saying they can't do it."

The challenge for Tony is to find and then nurture nurses who will be able to combine a high level of nursing knowledge with emotional maturity, sensitivity, awareness, diplomacy, energy, commitment, and excellent communication skills. Tony's ideal nurse is master of what he calls "both the art and the science of nursing. You can have a lot of theoretical knowledge, you can have a lot of technical skill, but if you don't have the people skills, you don't have the compassion, you'll be a great technician but you won't make that link with the patient that makes the difference. You have to be able to go into a patient's room and in a very brief period connect with the patient and do your physical assessment at the same time. But if the patient feels like you've given them your utmost attention during that brief time, they're going to feel better, and ultimately it's better for their healing process.

"Yes, you have to do blood pressure and the vital signs, assess the wounds. You need to know the anatomy and physiology behind what you're seeing: Why is the blood pressure dropping? Why is it when we give this medication, this happens? But if you're not talking to the patient while you're doing it, if you're not working with the family, supporting them, you're a good technician but you're not a good nurse. The technical things you can learn, but the compassion and the caring has to be innate. With this patient population, if you can't give to the patients and family emotionally, you're not going to last. It is taxing. And so, in some respects, it's self-selecting. In the end, we get the staff that is able to do the job."

A dapper man with neatly combed salt-and-pepper hair and a matching mustache, Tony is universally acknowledged to be a terrific boss. He wins particular praise from nurses with young children, who are grateful for his support as they try to balance the demands of job and family. In return, he has earned their loyalty and willingness to go the extra mile. From Tony's point of view, keeping nurses happy is the best bet that he will be able to keep them, period. It is the return on the considerable investment he makes in staff development.

Any nurse Tony hires goes through a multistep orientation period that begins out on the floor and progresses to the unit. New hires are paired with experienced nurses who act as preceptors, instructing, then observing and supervising until the neophyte attains a level of comfort and competence. It could take six months; it could take two years. "The orientation manual is this thick," Tony says, holding his hands apart like a fisherman sizing the one that got away. "Close to the end of the orientation, more and more is shifted to the orientee. The preceptor is still there but steps back, letting the new nurse take on more responsibility."

Presiding over orientation is the other woman Tony leans on, Mary-Liz Bilodeau, who has a mile-long trail of letters after her name: RN, MS, CCRN, CCNS, CS, BC. As clinical nurse specialist, she is responsible for staff development and education, patient management, research, and consultation. She advises other hospitals and other MGH departments on managing burns, does community outreach and public education, and presents lectures and workshops at national and regional burn and critical care conferences. You see her in business attire, in her office, on the phone, on her way to a meeting, all crisp manner and ice-blue eyes, but you also see her in scrubs at the bedside, working with the nurses, teaching, troubleshooting. She wears a white coat on clinic days, when in her nurse practitioner role she sees outpatients on a par with the attendings, managing patients, making referrals, writing prescriptions, "in the same way that physicians do," she said.

Mary-Liz is one of the new generation of "hyperprofessional" nurses. Advanced practice nursing is an answer to the age-old stereotype of the nurse as the doctor's handmaiden, and often the doctor's wife. That answer raises its own question, however. Is being more like a doctor the only way for nurses to get more respect? Bedside nurses vigorously refute that, as well as the hoary assumption that a nurse is someone who does not have what it takes to be a doctor. The two roles are different, and distinctly so. The mantra goes: Doctors cure; nurses care. Within limits, advanced practice nurses offer the best of both worlds.

Nurses speak of "the bedside" as an almost sacred place. It is their place, of course, the place where patient care happens, versus the OR, which is the doctors' domain. It would be nice to say Bigelow 13 is free from politics, but it is a piece of turf like any other workplace. Although what is best for the patient is always in the forefront, there is more than a little static in the background. Mary-Liz knew that coming in. "Nurses are a hard sell for advanced practice roles, because they're focused on their care at the bedside, and they want to know, 'What is it you're going to do for me? How are you going to make my life better?'" In her view, she can do a lot. "I can help them problem solve at the bedside. Because I'm credentialed, I can write the orders for them, so when nobody else is around they don't have to wait or call for a doctor, and we can move things along pretty quickly. I can also assist with their professional development and advocate for them." And she does.

Still, it is hard to ignore the swash as Mary-Liz sails into the unit on a mission and the backwash of muttering left in her wake. A patient is in trouble six ways to Sunday. All of his wounds have been covered by grafts and he has been moved into room 24. But he has developed melting graft syndrome—his grafts are not taking, his skin is breaking down, he is a magnet for infection, the steady progress that got him out of the unit and onto the floor has been halted, and if something is not done soon, he may find himself back in a BCNU. Mary-Liz thinks it is worth trying a special bed that is designed to reduce shearing and pressure on his grafts and might get him back on the road to healing. She asks a group hanging out at the back of the unit what they think about replacing the usual air-filled pneumatic beds with a Clinitron, which is stuffed with millions of tiny ceramic beads suspended in a steady flow of air. She is floating the idea, getting the opinion of various members of the team. She knows not everyone is likely to agree, but she wants to get everyone thinking about the problem. The physical therapist is concerned it will set back his work with the patient because optimal positioning will be impossible and stretching limbs to maintain range of motion will be extremely difficult to accomplish with the patient lying on the superbuoyant surface. Dressing changes will be brutal for the nurses, someone observes. The bed is famously uncomfortable for the patient, someone else notes. The respiratory therapist is less than enthusiastic because the head of the bed cannot be elevated, but the patient also has breathing problems that will be worsened if he must lie flat. Propping him up on pillows or a foam wedge will expose his grafts to the very shearing forces the bed is intended to eliminate. It sounds an awful lot like a bad idea, one might even think it is a dead issue.

The next thing you know, Mary-Liz is bringing it up at multidisciplinary rounds, the weekly Tuesday-morning meeting where the entire team—attendings, rehabilitation and respiratory therapists, psychiatrist, social worker, case manager, dietitian, pastoral counselor, consultants from infectious disease, pulmonary, ophthalmology, neurology, whatever discipline is relevant to the case at hand—crowd around the oval conference table to discuss each case. At the center of the meeting is the senior resident, who presents to the attending staff. Others join the colloquy as appropriate. Nurses are expected to participate when their particular patient is being discussed. They answer questions and sometimes give opinions or raise issues of concern to them or the patient's family. A plan of action is developed for each new patient; discharge plans are discussed for patients on the mend; existing treatment plans may be modified in light of complications that have arisen. It might have to do with a change of pain medication or diet, more surgery, further investigation, a family intervention, any of the vast range of treatment issues that invariably surround a burn patient. It may have to do with ordering a special bed.

A few days later, the Clinitron bed is in room 24 and the patient is in the bed. The consensus of the multidisciplinary team was to give it a two-week trial. But one person's consensus is another person's gripefest. During one of the informal break time chats that take place through-out the day and night, a clearly disgruntled group sits around the oval table.

"I think he's going to lose a lot," the physical therapist says. "I don't have great experience positioning someone to stretch in that bed because it's got too much give," yet turning off the airflow during PT sessions would make the bed like "a slab of concrete."

Someone says the patient looks like a turtle flailing on his back. Some-one else brings up the patient's psychological well-being. "He's trapped in bed for two weeks, and he *likes* getting out of bed."

"Right. People tell me that those beds are not comfortable. They're very warm, they feel claustrophobic."

Not surprisingly, Mike Wilson is acting as mediator, doing his best to soothe ruffled feathers and move things forward. He explains that every-one shares the concern about how much hard-won muscle tone and range of motion the patient will lose. But unless the grafts heal, all the muscle tone in the world will not make a difference. "It was a judgment call, but it wasn't an isolated decision," he says.

Finally, Mike says, "The whole thing is going to boil down to how we treat him. No matter what kind of bed he's in, it's going to have a lot to do with how cognizant we are that his grafts are very fragile." It is nurs-ing's responsibility to handle him in such a way that they do not shear off, Mike says, urging everyone to think of it as a learning opportunity.

That seems to do the trick. The PT gives a resigned smile. "Okay. You know what? I'm here over the weekend. I'll watch him closely, make sure that things are going as they should be."

The truth about burns is that easy answers are in short supply. There will always be competing and even conflicting priorities. Pain meds may worsen breathing problems, breathing tubes may ulcerate a smoke-injured trachea and feeding tubes erode delicate nasal structures, steroids may be the best answer to respiratory distress but a terrible idea for a pa-tient whose immune system is already compromised, replacement fluid may be necessary to prevent shock but can worsen edema, grafted limbs must be immobilized but immobility may result in loss of function. Burns are just about the most complicated medical disaster that can be-fall a person. Every member of the burn team has a stake, but a balance must be struck. The surgeon managing the case is the final arbiter, but he or she relies on input from everyone else.

For their part, nurses can give as good as they get. An ophthalmology fellow comes into the unit, called up to check a patient's corneas. She approaches the oval table, where a small group of nurses and one or two therapists is taking a break. Maybe there was something about the way she asked about the patient, but she gets nothing but a shoulder shrug and an "I don't know. You'll have to ask the senior." Who is, as it happens, not on the floor, and the nurses cannot say where he might be. In fact, there is little that nurses do not know, and most of the time they are happy to share.

Jen Verbesey had an excellent working relationship with the nurses on Bigelow 13, but she knew it was up to her to get it right. "I think it's personality a lot; some personalities tend to sometimes have some friction with the nurses, but I found that if you use them as a resource and clearly let them know that you understand that they've been there a lot longer than you have and know a lot more than you do, then you'll have a great relationship with them. If something happened, my first question to the nurses would always be, 'What do you do in this situation?' And nine out of ten times they could tell you exactly what's usually done. Obviously with your medical knowledge you can decide if that's a reasonable course or not, but they can really help you out. And if they don't like you, they can really not help you out."

* * *

Frank Ireland sits on a high wooden stool, watching Amy, the young nurse he is shepherding through orientation, as she checks the meds she has brought back from the Omnicell, the computerized dispensary out in the corridor. The beige multidrawered console is a sort of electronic Automat that is stocked by the pharmacy department with the standard meds used for most patients and anything special the doctors have ordered for an individual one. Amy punched in a code and then opened five or six drawers, each about the size of a post office box. As she took what she needed, each withdrawal was recorded in the machine's memory to maintain inventory and create a record of all medications used. At the end, the machine printed out a receipt like the one you get at the self-checkouts at the supermarket. She brought everything back to the unit on a cafeteria tray: bottles, jars, ampules, and sealed plastic containers like the ones usually filled with jam and creamer and pancake syrup. Now she is verifying it all again, checking off the contents of her tray against the physician order entry printout (POE, pronounced *Po-ee* for short).

Their patient is an elderly man who had been burning leaves in his backyard when he tripped over a rake and fell into the smoldering pile. A younger person might have been able to jump right up, might not have fallen at all, but by the time this man managed to roll out of the way, he had sustained second-degree burns that ran in a wide stripe up his left side from his knees to his head.

Aside from the usual medications he needs as a burn patient, the man also has heart problems and has to continue on the drugs he was taking before he landed in the hospital. He inhaled quite a bit of smoke, a strain on lungs already at risk. His urine is scanty and nut brown, a sign of kidney-killing rhabdomyolysis, and one glance at his wasted frame confirms that he is catabolic, the condition exacerbated by inhalation injury. The liquid feed flowing from the enteral pump is not keeping pace with his pumped-up metabolism and the replacement fluid is not sufficient to dilute the overload of toxic waste in his urine. The lab reports will tell the full story and orders will definitely be changed accordingly when the attending and residents do their rounds. Amy is learning, and in time she will know all this by heart, be able to anticipate, maybe even, like some of the experienced nurses do, actually tell a new junior resident what they have done so the orders can be written retroactively. But for now she runs it all by Frank, who nods like a smiling Buddha and occasionally offers words of wisdom.

Having mastered the routine out on the floor, Amy is nearing the end of her critical care orientation in the unit, but it remains an open question whether she will make a career in burns.

Bigelow 13 has a mixture of young nurses—new to the profession, new to burns, and mostly on their way elsewhere—and older nurses who have been working with burn patients for years. Many of the latter are lifers who began their careers at the General and will end them there. And many of them are nearing retirement. The new generation comes in with a different set of goals and expectations. "People coming out of school now have definite two-, five-, ten-year plans," Tony DiGiovine said. He recognizes that those plans rarely include staying there for the rest of their lives.

Sally Morton and Mike Wilson are lifers. Bob Droste and Lynn Bellavia came over from White 12, and Mary Williams has been working on Bigelow 13 since it was a general surgical floor, long before burns arrived. Frank Ireland is on his way to being a lifer, as married to Bigelow 13 as he is to his wife and fellow burn nurse, Lisa. In fact, he says, "There are thirty other women here who think they're my wife, too. They have no reservations about bossing me around."

Frank is one of five male nurses on Bigelow 13, six if you count Tony. They are outnumbered by the women by about six to one. There is no question that nursing is largely women's work, with men representing about 5 percent across the profession, but burns is one of the areas that attract men, emergency and trauma being others. Perhaps it is because it is such hard work, punishing in its demands on both body and psyche, though you would get an argument on that from the legions of women who function at the highest level amid the sights and smells and physical intensity of burn nursing.

Rick MacDonald is one of the younger generation, and he is a man on the move. He is actually nearing forty, though he looks like he is in his late twenties. Nursing is a second career for him. He started out as a carpenter, self-employed and uninsured. "One day when I was banging nails up on a roof I decided I didn't want to be doing that when I was sixty years old." His girlfriend's mother was a nurse, and he thought it might be interesting work. Before taking the plunge, he took some EMT classes, just to get familiar with medical terminology and some basic concepts. Then he enrolled in nursing school, working as a nurse's aide at a small hospital until he graduated and then received his RN license in 1996. His first job was at the General, working on the neurology floor for three years and then as a floater, which gave him the opportunity to check out other areas of nursing. He liked what he saw when he was assigned to Bigelow 13. "I could tell this was a good group, always helping each other." He has been doing burns for a bit over a year, working both day and night shifts, and he expects to stay one more year. But there is no question Bigelow 13 is just a stop along his career path, one he hopes will lead to certification as a nurse anesthetist, an advanced practice specialty that requires one year of ICU experience and two years of additional school, full-time, to gain a master's degree. After that, he will look for a position in a dentist's office or outpatient surgical facility—day jobs, regular hours, better money. Rick is a married man with a young son, and he knows that the nursing shortage means he can pretty much pick and choose what he wants to do and where he wants to work.

Right now, where he is working is at D bed, and what he is doing is assessing his patient, a man whose left arm has been amputated above the elbow, so deep was the burn, so complete the destruction. The patient's chest is encased in heavy dressings, halfway through the staged grafting of his upper torso. The details of how he was burned are still murky. When he regains consciousness, perhaps he will have a story to tell. In the meantime, the police are investigating and Carla Cucinatti is trying to locate family or at least someone who knows this man, apparently homeless, probably a drug addict, most likely mentally ill. None of that makes a difference to the work Rick is doing.

Spread out on the portable counter that marks the invisible boundary between his work area and Amy's is a double-sided trifold sheet, the equivalent of six sides of 8½ by 11 paper. Each sheet represents a single day, midnight to midnight on Bigelow 13, with hourly readings of temperature, heart rate and rhythm, respiratory rate, blood pressure, oxygen saturation, and on and on. Every substance that goes into and comes out of the patient's body must be noted, all interventions recorded, neurological status evaluated, behavior, range of motion, level of pain, degree of sedation—every detail of this patient's complicated existence on this day, at this hour must be filled in and filed in a three-ring binder.

At Mass General, this documentation is done by hand, though some of the data is taken from the monitor, which retains a twenty-four-hour record of vital signs before its small memory is wiped clean. At other hospitals, all recordkeeping is computerized, and in those institutions whose information technology is closer to the cutting edge than it is at MGH, it is all centralized. On Bigelow 13, patient charts are a mishmash of handwritten notes and transcriptions of what the attendings dictate, orders typed into a computer by the residents or copied in from the POE by one of the operations associates, and the endless assessment sheets. In addition, nurses must "classify" their patients once a day, entering data into an entirely different system, Medicus, which helps the hospital administration track the allocation of personnel resources in relation to patient needs and measure nursing productivity against specific benchmarks. Nurses above all spend an inordinate amount of time documenting the enormous amount of work they do. Clerical work is no small part of the daily routine for these magicians of multitasking.

Rick adjusts his visor and pulls on the gauntlets, securing them at the top with rubber bands. He reaches into the BCNU, but before he begins his work, he places his hand gently on his patient's shoulder. "How you doing, Buddy?" he asks. "How're you feeling today?"

All the nurses do that, though each has an individual style. Some use their patient's first name, others will say Mr. or Mrs. So-and-So. Rick usually calls his patients "Buddy," even when he knows their name, whereas Sally Morton, who has children and grandchildren of her own, tends to use a motherly "Darlin'" to encourage and comfort her patients. The voices can always be heard, amid the beeps and clicks and chuffs of the unit, over the background music, the buzz from the oval table, the clarion call of the PA system. The nurses keep up a running dialogue, announcing what they are about to do, talking their way through whatever it may be: "I'm just going to wash your back now, Mrs. Smith." "I know you hate this, Buddy, but we have to do it. It'll just take another minute." "I'm just going to listen to your lungs, Sweetie." "You're doing great, my friend." "There you go, Darlin'. I'm all done now." And never will a nurse at bedside say anything she or he would not want the patient to hear. No one is certain whether these deeply sedated patients can hear what anyone is saying or whether they will remember their time in the unit. Chances are they will not, but the policy is to assume they might, so the nurses speak to them, calling out as though across a vast divide, ever aware that there is a person inside that mummy wrap, someone's father or sister or husband or child is lying in that bed, and that person has a history, a past, and, everyone hopes and works toward, a future.

CHAPTER EIGHT

Gasping for Breath

Shortly before seven o'clock each morning, Bigelow 13 starts to come alive. The morning nursing shift arrives, changing into their scrubs and stepping into their clogs, leaving the floor of the dressing room strewn with street shoes like a nightmare version of a teenager's bedroom. For fifteen minutes at least there are twice as many nurses on the floor as usual. At bedsides, in the corridor, squeezed into a corner of the front desk, sitting at the oval table at the back of the unit, sweaty and exhausted night nurses hand off their patients to bright-eyed, energetic replacements.

Carolyn Washington opens the door to her windowless office behind the front desk and greets her staff as they assume their duties. Two operations associates are at the front desk, one a young man with cornrows who looks like he could still be in high school. Carefully, deliberately, he transfers orders from the POE printout to medication sheets that the nurses will refer to throughout the day. Phones start ringing, the PA system punching through the general hum of the floor. A young unit service associate pushes a stainless steel cart into the unit and begins restocking shelves with supplies the staff went through like locusts the day before: blue paper–wrapped, autoclaved packages of dressings and instruments, boxes of latex gloves, paper masks and hats, yellow plastic aprons, bottles of sterile saline. A faint smell of toast and coffee floats above the indescribable but unmistakable mix of antiseptic solutions, soggy bandages, secretions of various sorts, a sourish odor that stops just short of sickening and is as imperceptible to those who work here as it is assaultive to a visitor. Breakfast is for floor patients who are able to eat solid food. For the rest, it is indistinguishable from every other meal, either the nutrient-rich formula that flows constantly through a nasogastric tube or a gastrostomy,

which runs directly into the stomach through an opening in the abdomen, or for the few who cannot tolerate enteral feeding, total par-
enteral nutrition, which drips directly into their veins. Many patients also
get replacement fluid through an IV line. Fluid in, fluid out, all metered
and measured and carefully noted by the nurses at regular intervals
throughout the day. The burn techs ready their supplies for the morning
dressing changes, then check in on the floor patients, take their vital signs,
record their I and O (input and output, which is a polite way of referring
to the business of bedpans and urine collection bags), bathe those who
need help, feed those who are on solids but cannot manage by themselves,
talk to them to ease their loneliness, calm their fears, and if possible lift
their spirits.

The oval table is littered with paper coffee cups, half-eaten donuts,
crumpled wax paper, amidst a collection of three-ring binders filled with
patient charts. A resident sits at a computer near the window, her back to
the table, finishing up the day's orders. Six or so nurses sit around the
table, paired off in businesslike tête-à-têtes. The conversational murmur
plays against the usual sounds of the unit, a rare bit of easy listening music in deference to the early hour, the chuff of the respirators, and the
monitors beeping a madman's syncopation outside the four tented beds.

Nancy Giese is one of the nurses on the way in, sharing with her night
shift counterpart an overlapping half hour for the handoff, what nurses
call report. Nancy drapes her lanky form into a chair, within earshot but
with her back to her patient's bed as the night nurse gives a full oral report. A tall woman with long, expressive fingers and mesmerizing hazel
eyes, Nancy exudes a calm confidence that marks her as one of the more
senior nurses. She came to the General fresh out of nursing school and has
been on Bigelow 13 since the late 1980s. "To be totally frank, I took the
job at Mass General as a new grad to get one year's worth of experience so
it would look good on my résumé, because if you've worked at Mass General, you can work anywhere. People think that you're something special.
Well, it's sixteen years later and I'm still here. So I think that says a lot. I'm
a firm believer that if it works you should keep doing it, and when it stops
working, you try something else. It still works for me."

It works because the nurses are together in this single large room and
can summon each other for assistance or a few minutes' coverage while
they check on a floor patient, take a phone call, or make a food run for

the entire group. The general pattern is that the most experienced nurses are assigned one patient in the unit and one out on the floor, the burn techs helping with dressing changes and doing the lion's share of the hands-on care for the floor patients. Some patients are so ill that they require a nurse's undivided attention, around the clock. And sometimes things get so intense in the unit that a nurse will call for help from anyone who can leave the bedside to lend a hand.

When there is a crisis, no one is left to handle it alone. "Depending upon what else is going on, everybody comes to your rescue and does a little piece of this and a little piece of that, because they're taking care of their patient too, but we're all in that position at some point where you need all the help you can get, because you need to be doing about ten things at one time and you can't. And that's why this place is wonderful. We don't have that—nyah-nyah—'well, you didn't help me last time.' You know, the junior high attitude doesn't fly around here. The work's too heavy. So people will come over and help you out. And you don't have to go and ask, to beg. It's just a natural—you need help, so people come over and they stay as long as they can until their patient needs something. If you start to get behind, you can say, 'Hey, Frank, can you give me a hand with this for a second,' or 'Lynn, can you check this for me because I'm having a really hard time, I can't get it, I need some help here.' And it's not a big deal, I mean, it's expected that that's how you work together."

Nancy thinks Bigelow 13 is unique in that way. "It's a family. Other units don't have that, or not to the same degree. You know, we're not real pretty up here. I mean, we're *beautiful,* but we all smell really bad by the end of the day, all the women have makeup that is, well, you wouldn't go out in public like that. We all wear pajamas to work; they're ill fitting, and we don't care! We don't have a nice clean job. We're doing nasty things—it smells horrible; it's messy; it's dirty; it's very physical; it's heavy. And so we're all in the same boat together and we all have to help each other because you can't do it yourself."

The best example of the teamwork can be seen when a new admit arrives. One nurse will be assigned to take care of that patient, but for the first several hours at least it will be more than a single person can manage. The primary nurse is in charge, but all the others do whatever they are asked, so long as it does not compromise the needs of their own patients. That usually means giving five minutes here, ten there, completing one task or fetching one piece of equipment or sending a specimen down to the lab by way of the pneumatic chute in the wall outside Mary-Liz's office or going to the front desk to fetch the printout of results sent back by the lab or any of the myriad ways they can lighten the load.

"Everybody helps," Nancy said. "You work together as a team, so it goes quickly. If it takes too long, then bad things happen."

The first thing on the agenda is getting the patient's temperature up to normal. It is not unusual, Mike Wilson explained, for a patient to arrive in the unit with a temperature in the low nineties. The longer the patient has been down in Emergency, Bob Droste said, the more likely this is to be the case. Even the warm room in the ED cannot achieve or maintain the high levels of heat and humidity these patients require. "The space isn't very warm, the patient's exposed during all this time. They have fluid warmers, but it's just not enough," Bob said. What they do not have is BCNUs, which can be stoked up to 100 degrees if need be. So they try to get burn patients upstairs as fast as possible. Sometimes possible is not very fast. Dan O'Shea had to go to the HBO center first and then wait until a bed was made available. When the unit is at capacity and the floor close to full, finding space can take hours. "If that's the case, somebody in Emergency has to keep them going, resuscitating them, checking for pulses, doing all that stuff down there. And they know how, but it's very difficult for them if they're all backed up, because it takes a lot of personnel," Bob explained.

Once the wounds have been assessed and a thorough burn diagram filled in, the nurse puts on the dressings. If the patient was wrapped downstairs, revision may be needed. Burn nurses know how to bandage a burn patient by heart but others tend not to. For example, the hands have to be exposed so a nurse can check pulses to make sure blood is flowing to the extremities.

"So on top of running the fluids and doing the dressings and checking the vent, and suctioning the patient and doing all that, the nurse is checking for pulses," Bob said.

"And checking just about every electrolyte in the body," Mike added. "You're sending lots of labs off: potassium, magnesium, calcium. You send blood samples off, you get them back, and then you replace whatever needs to be replaced."

The nurse appears to be running the show, and it is true that nurses on Bigelow 13 have an enormous amount of autonomy. But they are acting on written orders from an MD, usually a resident. Ideally, the orders are spelled out according to protocol.

"An experienced nurse, acting out of experience, will know what to do," Mike said. "You do what you need to do to save somebody's life."

* * *

Sally Morton has gotten report at the bedside and she is already at work suctioning out her patient's airway, a regular task with all patients on mechanical ventilators. They are lying still and cannot cough out the secretions that continually collect in their airways and threaten to gum up the works. Sometimes the nurse does it, sometimes the respiratory therapist. It mostly depends on which one has the time and what else is going on. Sally likes to be the one who does the first suctioning of the morning: to get a sputum specimen to send to the lab, to see how copious and how thick the secretions are, to get a handle on how often she will need to suction during the day.

She slides a line of hosing into a port at the side of the endotracheal tube. "Here you go, Connor, going down. It feels like I'm going to take your breath. Yup, starting to tickle now. There you go. I'm going to take away your breath now." It sounds like the device that slurps up saliva during a dental procedure, sucking out sputum and anything else that might be blocking the passageway to the lungs. Sally will do it three or four times, giving her patient a break between each pull. Even though he is unconscious, he gags and his chest heaves with each suck, a paroxysmal response to a sensation no conscious person would be able to endure.

"Okay, Darlin', good cough, great cough. Wonderful, wonderful. We're going to clean that out now." The cough is reflexive, the encouraging words unheard.

On the next break, Sally brushes his teeth, soaking a small, bright green sponge on a plastic handle with chlorahexadine to replace the saliva bath and constant swallowing that keeps a conscious and healthy person's mouth reasonably clean.

"Here you go, Darlin'. Just brushing your teeth here. Just like you do at home. It's *so* important. And it looks like you've got beautiful teeth.

You want to keep those teeth. Here we go. Now I'll get the back side of your teeth. I know you don't like it." He clamps down on the sponge, another reflex. Sally leaves it there until his jaw relaxes. "Sometimes the better part of valor is just to leave it in there until he's nice and calm and then just take it out. I want to be sure I come out with the whole sponge," she explains.

"Okay, Darlin', I'm going to do it one more time, cleaning out your lungs. One more time, take away your breath. There you go. Boy, we needed to clean this out, didn't we."

Through it all, the monitor emits a steady *beep-beep, beep-beep*. It is the second most serious of the three-tone repertoire. Sally knows it is the patient's central venous pressure responding to the suctioning, but she checks the screen anyway. The *deet-deet-deet* heard all day long is fairly routine; it usually signals that a lead has detached or lines have gotten tangled. The real heart stopper—literally—is a *bing-bing-bing* that the staff translates as "I am dead." Such is the black humor that keeps the staff sane.

* * *

You do not have to inhale carbon monoxide to end up on a respirator. You do not have to inhale hot air. All you have to do is spend enough time in a smoky room to sustain life-threatening damage to your airway.

When Drs. Finland, Davidson, and Levenson studied respiratory damage in the Cocoanut Grove patients, they made the connection between time of exposure and severity of injury, but what they did not know is how smoke does its damage.

The respiratory tract begins at the nose and mouth and ends in the lungs. In between lie the upper tract—the pharynx, or throat, and the larynx, or voice box, which contains the vocal cords—and the lower tract—the trachea, or windpipe, which branches at the lungs into the left and right main bronchus, which in turn branch into five smaller bronchi, one each for the two left lobes and three right lobes of the lung. Inside the lungs, the five bronchi branch into hundreds of bronchioles, which branch into millions of alveolar ducts ending in alveoli, tiny sacs that cluster like grapes at the end of each duct. There are an astonishing 300 million alveoli in the average adult lung, with a combined surface area averaging 1,000 square feet. The exquisitely thin alveolar walls are embraced by a network of equally thin-walled capillaries. It is here that the life-sustaining gas exchange takes place: oxygen and nitrogen pass from the alveoli through the capillary walls and into the blood; carbon dioxide and other waste gases move in the other direction.

Everything above the bronchi, from the nose and mouth on down, serves as air conditioner and conduit, moisturizing and tempering air as it travels to the lungs. Air is an extremely poor conductor of heat, and the upper airway is a very effective cooling duct. By the time air gets to the vocal cords, it is usually at body temperature. That works whether the air is too cold or too hot to start with.

It is possible to sustain thermal injury of the lower respiratory tract, but it is quite rare. An industrial accident involving exposure to pressurized steam, which has 400 times the heat-carrying capacity of air, can transfer a huge amount of heat into the lungs. Flame will burn anything it touches, even if it is below the larynx. Case manager Pauline Buskiewicz told the story of a circus fire-eater who came to clinic once with burns halfway down to his lungs.

The overwhelming majority of fire-related inhalation injuries, however, are chemical rather than thermal burns. The airway cannot filter out the products of incomplete combustion—the acids and alkalis contained in smoke and soot. Those substances set off a chain reaction that shreds the lining of the airway. Like the fluid shift, it results from a combination of cellular changes and an extreme inflammatory response.

The airway is lined from top to bottom with mucous membrane, an open-faced sandwich of basement membrane topped with epithelial cells that are distant relatives of the cell type that makes up the epidermis. The chemicals in smoke dissolve the "glue" that attaches the epithelial cells to the basement membrane and the cells peel off. It is the same thing that happens to the roof of your mouth when you put too much red pepper on your slice of pizza. It is not so much the molten cheese as it is the chemical capsaicin in the pepper that strips the cellular layer from the mucous membranes lining your mouth. Imagine that extending beyond your palate, all the way down the airway and into the lungs. Like the hyperpermeable capillaries, the naked basement membrane is a protein sieve. The protein that oozes through it combines with the epithelial cells to form a sticky mass that clogs the bronchi and bronchioles, narrow passageways that are further narrowed by swelling due to the inevitable inflammatory response.

The gummy mess forms into casts. John Schulz said they look "just like little branching trees," cast in the shape of the tubes they fill. Patients who are not intubated may be able to cough them up. Those who cannot cough need to be suctioned. The moist air coming from the ventilator helps loosen the secretions, as do chest percussion and changing the patient's position, especially the rolling from side to side that happens during dressing changes. The beds in the unit even have a vibrating feature reminiscent of old-time motel beds. The manufacturer claims this helps to jiggle the gunk loose and make it easier to suck out, a claim Mary-Liz Bilodeau disputes. But positioning, chest pounding, and blind endotracheal suctioning of the sort Sally and the other nurses do several times a day can go only so far and do only so much. When the casts are thick, numerous, and lodged deep in the bronchi, they pose a significant risk. Even if they do not totally block the airway, they are a breeding ground for bacteria and a sure route to pneumonia. The best way to find and remove bronchial casts is with a bronchoscope. A physician—usually a resident, sometimes one of the attendings—will periodically "bronch" any patient with inhalation injury or a lung infection. It is a bedside procedure, but it is done under anesthesia.

Respiratory injury is progressive. The "pizza peel" happens almost immediately after exposure to smoke. About three hours later, polymorphonuclear neutrophils, those hyperactive immune cells, start gobbling up dead cells. The free radicals released in the process attack the delicate alveoli, inflicting damage and severely compromising gas exchange. It can take a few days for respiratory symptoms to emerge, but it is so important to treat inhalation injury early, before severe and irreparable damage is done to the lungs, that everyone keeps an eye out for early signs. The first may be apparent immediately: face burns and soot in and around the mouth and in the sputum. Even though the burns on Tom Parent's face were superficial, they were a sign that the fire was close enough that he may have inhaled smoke. That is why they watched him for the first few days, or rather listened, for squeaky, creaky, or crackly breathing, breathlessness, and a cough.

Fortunately, Tom never developed any breathing problems. Dan O'Shea, on the other hand, got out of the hyperbaric chamber and up to

Bigelow 13 just in time to develop all the signs, symptoms, and complications of full-blown inhalation injury.

* * *

It was not just the carbon dioxide. Dan had spent a long time in a very smoky room, first asleep, then unconscious, breathing in smoke all the while. As bad as Dan's burns were, "the hallmark of his admission was his pulmonary problems. His early course was dominated by worries about that. He was on toxic ventilator settings, basically, because of his lung injury," John Schulz said, referring to the high percentage of oxygen they were pumping into his lungs. During the first week he was in the hospital, he had bronchoscopies on average every two days. Each time, large bronchial casts were removed, the first few blackened with carbon that had made its way down to his lower respiratory tract. "His lungs rapidly got worse over the first four to five days," Schulz recalled. And then he developed pneumonia.

If was not so much a matter of *if* as it was of *when.* "Almost everybody who has an inhalation injury gets pneumonia," Schulz said. "Between the immune compromise and the wrecked state of his lungs," Dan was ripe for it. "He has many kinds of bacteria living in him and on him. He's not coughing and doesn't have the usual things to get rid of bacteria. The mucosal barrier is broken, so anything that's remotely pathogenic will just hop on. The BCNUs are really designed to prevent cross transmission, but you can't keep these people from getting infected."

One day, about two weeks in, Dan's temperature shot up to 105. When they cultured his sputum, it came back positive for methicillin-resistant *Staphylococcus aureus,* MRSA. *S. aureus* is one of the most common bacteria found on the skin. You probably have it, I probably have it, and it is not going to make either of us sick. Our intact skin keeps it outside, and should it sneak in through a cut, our strong immune systems will zap it before it gets very far. Burn patients are not so lucky. When a burn patient develops an MRSA infection, it represents a danger for all the other burn patients in the vicinity. For that reason, MRSA patients must be kept in strict isolation. On Bigelow 13, the BCNUs are, in effect, freestanding isolation rooms within the ICU. In other burn centers patients usually occupy separate rooms, and MRSA patients always do.

In either case, the highest level of precautions must be taken: everyone who comes in contact with the patient must be gowned, gloved, and masked, and the protective coverings removed before they leave the room or, in the case of the BCNU, the bedside. When the patient vacates the room or BCNU, it must be thoroughly disinfected—floors, walls, ceilings, windows, curtains, every surface must be thoroughly scrubbed.

S. aureus also happens to be very good at developing resistance to the antibiotics commonly used to treat it, methicillin and other drugs in the penicillin family. MRSA infections are usually treated with vancomycin, a very powerful antibiotic that is able to defeat most of the defenses of resistant bacteria. It must be given intravenously, and in a slow drip, or it can cause a sudden drop in blood pressure. It also plays havoc with the metabolism and can damage auditory nerves, causing ringing in the ears and sometimes deafness. It is not a drug to mess with unless absolutely necessary, but it can be a lifesaver. Unfortunately, some strains of bacteria have developed resistance to vancomycin as well. Medical science is running out of antibiotics to fight resistant bacteria.

By the middle of February, it was clear that Dan was going to be on the respirator for a long time. It was equally clear that leaving the endotracheal tube in place much longer was not a good idea. For one thing, intubation requires heavy sedation. "It's a noxious thing to have that tube sticking down in your airway, without sedation," John Schulz told me, so patients are put into a sort of drug-induced coma, and many are given paralytic drugs so that the chest muscles are completely relaxed to allow the respirator to do all the work of breathing.

Using strong sedatives to induce a coma—what some in the unit call "snowing" the patient—may be done without inducing paralysis, but never vice versa. "People that were paralyzed without sedation compared it to being buried alive," Mike Wilson told me. "We don't do that." The pain, sedation, and paralysis protocol, he said, "has written into it ways that we check not just the level of paralysis but the level of sedation. We have ways of looking at them individually as well as collectively. There's a lot of safeguards in our policies to prevent that."

In addition to requiring heavy-duty sedatives, the ET tube may over time erode the delicate tissues of the mouth and throat, exposing another area of the body to infection. "If we think someone's going to be ventilator dependent for two or three weeks, he needs a tracheostomy," Schulz said.

A tracheostomy is a hole made in the throat at the level of the trachea into which a tube is inserted. Not only is it more comfortable than an ET tube, but it makes the transition off the ventilator much safer. Instead of removing the entire apparatus, and possibly having to put it back, as happens with an ET tube, the connection to the vent can be taken off and replaced with ease and, if the patient is having trouble breathing without assistance, speed.

When John Schulz explained all that to Peggy and Jack, they found it hard to believe that cutting a hole in Dan's neck near his Adam's apple was less of a deal than a tube down his throat, but he assured them that it was a step in the direction of getting Dan off the powerful sedatives that were keeping him unconscious and off the respirator entirely. At the same time and for a similar reason, they planned to replace the nasogastric tube that was carrying liquid formula to his stomach by way of his nose with a percutaneous gastrostomy, or PEG, a tube that went from a hole in his side directly into his stomach. It sounded horrible to the O'Sheas, but according to Dr. Schulz, when the time came to remove the tubes, both holes would close by themselves, without needing stitches or anything else. It was like a splinter, he said. Once you pull it out, the skin closes up in a few days.

So the next time Dan went into the OR, he came back with a tube in his throat and a tube in his left side. For the first time in two weeks, his parents could see his face without a tangle of plumbing in the way.

* * *

Caring for a patient with inhalation injury is a full-time job that cannot be done without a respiratory therapist. Mike Marley is the man who keeps the vents running on Bigelow 13.

Marley, "as in Marley's ghost," he says. He comes by the work honestly. As a child he suffered severe asthma, so respiratory therapists were his heroes. Most kids do not even know what a respiratory therapist is, but Mike knew he wanted to be one when he grew up.

He has been at the General for three and a half years, working the seven-to-seven shift. Another therapist covers the overnight. On his time off, Marley moonlights as a disc jockey or is at home with the Canadian lynx and the ocelot he saved from being made into coats. He claims they spend the time he is at work hanging out in the living room watching *Animal Planet.*

Mike is a funny guy who fits right in with the bonhomie of the unit, though he admits he was horrified by burns when he started the job. He got over it fairly quickly, though, thanks to the people he works with.

"I don't know why, but it's so different from the other burn units I've been exposed to. It's completely different, completely. The first burn unit I worked in was down the end of a hall and in a little kitty corner, and nobody talked about it, and even the nurses were horrified by it. Then I came here and it was just so much better, like night and day."

It looks like Mike Marley is lifer material. "Oh yeah, I love it. I'm really here because of the nurses. I've been in situations with people that I don't like working with and it's been absolute hell. Here, the doctors, nurses, everything just so completely clicked here. We have a lot of leeway here. Whereas most other hospitals you're told what to do, here we tell them what works, why don't we try this, and they're open in this place more than any other place I've worked. They like your input, they like it when you're proactive. There's so much more respect here. I've worked at hospitals where they've told us: You're one step up from housekeeping."

Even though he spends less time with each patient than the nurses do, he still makes a connection, taking the time to look at the photos family members bring in and the nurses tape around the BCNU before he starts suctioning out the airway of the person who lies inside or checking to see why the vent alarm is sounding.

This particular morning, Bob Droste's patient has been setting off the alarm nearly nonstop. His "sats"—the oxygen saturation in his blood, an important indicator of how effectively gas exchange is taking place at the alveolar-capillary border—dropped precipitously for reasons that are being investigated but are still unclear. It could be any number of things, according to Bob, from a faulty probe to a pulmonary embolism. "There was an acute, almost instant change this morning," Bob reports to Mike. "He was on an oxygen setting of 40 percent and the saturation was in the upper nineties and everything was going fine, and within minutes the sats dropped way down." Regardless of the cause, job one is to get more oxygen into his system. Mike begins to up the oxygen settings on the vent. Nurses cannot change vent settings; it takes a respiratory therapist to do that. And strictly speaking, it is done on a doctor's orders. The thing is, the only doctor on the floor is the junior resident, and he is as green as his scrubs, less than a week into his rotation. So Bob and Mike do what needs to be done, and the resident plays catch up.

There is a myriad of settings on the vent, but the one they are working with is the fractional inspired oxygen (FIO_2), the richness of the mix flowing from the vent into the man's lungs. It can range from 21 percent—the same as room air—to 100 percent, though too much time at the high end risks oxygen toxicity and a cascade of negative pulmonary consequences. They got it up to 70 percent, but the sats were still in the low 90s. The saturation goal is 100 percent—the percentage of binding sites on each hemoglobin molecule occupied by an oxygen molecule, most decidedly not the percentage of oxygen coming from the vent. They end up going all the way to 100 percent oxygen, a setting Mike calls "hideous," and finally manage to push the sats up to 98.

The information about the sats is coming from a probe clipped to the man's toe, and there is a chance it was giving bad data, so Bob has sent a vial of blood drawn from the A-line down to the lab to get an arterial blood gas reading. Blood gases are a complicated set of values that reflect both respiratory and circulatory function, and add up to how much oxygen is available to the tissues of the body.

Mike goes to get the lab results. When he comes back with the printout in his hand, he shoots a bunch of numbers at Bob. The bottom line: Things are much improved, but why the problem arose and then resolved has both Bob and Mike scratching their heads.

"Does it make sense to you?" Mike is laughing.

Bob deadpans a no.

"It's really weird. It's like an *X-File* going on over here. We've tapped into a vortex from beyond."

"Hell," Bob says, "I like it anyway. It's getting better and I don't know why. It's not that I don't care. I'd like to know why, but all I really care about is that this is much better, remarkably better."

Mike starts to turn down the vent, with Bob coaching him: "Gently, please, gently. I'm not in a hurry."

"I've got it down to 80. I'm going to go slow."

"Do we have an order for this?" Bob asks no one in particular, then answers himself. "I don't think so."

As if on cue, the junior resident comes by. "I heard we're going to go down on his vent again," the slender young man says.

"Yeah, we've gone down a little bit already. We never had an order to go up. You never actually wrote anything, so we just went up and now we're headed down. It would be sort of a good idea if the orders caught up with us." Bob's voice is completely neutral.

The resident pauses, then says, "I'll be honest with you: I don't know how to . . . I always try to do the orders on the vent and I don't know how to do them. You guys went up to 10, then 20, then . . . "

"What we did was we went from 40 percent to 70 to 100."

"What's he on right now?"

Mike is at the vent, on the other side of the BCNU. He calls over, "Eighty."

"Eighty percent?"

The exchange is a classic; the saving grace is that the resident knows what he does not know and is not afraid to admit it. He knows where to get the answers and neither Bob nor Mike is making him pay.

As Mike explained later, "The residents get a couple of weeks of ventilator management in medical school, whereas I had two and a half years of it in school and eight years' experience doing nothing but this. They *like* the fact they don't really have to worry about the vents unless there's a problem."

After a bit of clarification of the various vent settings—not just oxygen level but also the volume of gas, the pressure at which it is delivered, the number of breaths per minute, and the rhythm of those breaths—the resident smiles, finally up to speed. "Okay, I can write that."

"You can even make it simpler and write 'increase FIO_2 to 70 or 80.' I don't care, whatever you want." It is not that Bob does not care—he cares a lot. It is just that he has minimal patience for the niceties. His energy goes toward his patient; he has none left for unit politics.

* * *

There are no formal visiting hours on Bigelow 13. Families are welcome whenever they can come, unless there is a crisis or their loved one is undergoing a bedside procedure—from dressing change to bronchoscopy, things an anxious family member should be spared the sight of. Then they are asked to wait outside in the family area, the faux living room bridging the Bigelow and Blake buildings. The wait could stretch for hours. Jack and Peggy O'Shea came every day, and it was not unusual for them to arrive in the early afternoon—around two o'clock, say—and not get in to see Dan until after six.

It did not take long for Jack and Peggy to come to know the unit like it was their own backyard, and the people who spent time there—patients, families, staff—like they were neighbors. It was during those hours waiting that Jack became the unofficial mayor of Bigelow 13. He's that kind of person. He calmed a lot of nerves and quelled some major discord. There was one man they remember in particular. His wife was very badly burned and he was understandably anxious, but in Jack's opinion he was looking in the wrong place for a scapegoat. "He was really wearing on the nurses and everybody else, telling them all what to do." He even got angry at Mary, Peggy recalled. Jack shook his head at the notion. "But I said to him, 'Lookit, our son's been here for a while, and believe me, everything that they do is in the best interests of your wife.'" Jack managed to calm him down, but Peggy said that is when she knew they had been there too long.

They were encouraged to call at any hour. Sometimes Jack called at three o'clock in the morning, when he could not sleep and was worried about how Dan was doing. "I'm sure we were a pain in the ass sometimes," Peggy acknowledged.

Again, they finished each other's sentences, filled in each other's memories, as though the pain was still too raw for either of them to bear individually.

"Well, you're so uptight and nervous and protective and . . ." Jack offered.

". . . and the worst thing is you can't do anything, you *cannot* do anything. Like Mary said, you can reach in after a while, if you put the gloves on you can reach in . . . "

". . . and hold his hand. I knew he knew we were there . . . I know he did."

Even though they think the staff tired of them after a while—they were assertive in advocating for their son, they asked a lot of questions, they said what was on their minds—the O'Sheas are remembered fondly, and not just because Jack donated the four wooden stools that stand outside the BCNUs so the nurses can occasionally sit down. The high stools are showing their age, held together with duct tape, but no one wants to replace them. Bob Droste wiggled the one at his station. "I've been meaning to bring in some wood glue and a clamp and work on this one," he said, though it is hard to imagine how he would carve out the time.

Mary Williams thinks of the O'Sheas every time she sits on one of the stools. "They were great parents, great people. They loved their son, they were here every day. They were there for their son, regardless of what happened, they were always there. We will always remember that family."

And they will never forget Mary.

It was she who suggested they bring in a photograph of Dan, the way he was in happier times. "She said, 'We see them the same as you see them, all bandaged up. We don't know what they're like or anything about them.' So we brought a picture of the whole family. It was right there beside his bed, so that when they did things to him, they would say, 'Hi, Dan, we're going to do this now, we're going to do that now, and look who's here, your mother.' Even though he was not aware and he still does not remember any of this, you know, it just made it that he seemed like a person, not a mummy."

One day Peggy was sitting at Dan's bedside, gently stroking his bandaged arm and speaking softly to him. He had not yet opened his eyes, his breathing was still controlled by the respirator, he was still deeply sedated. But Peggy spoke to him and prayed for him just as she would if he could hear her. Suddenly the alarm on the monitor went wild. The bells and beeps always made Peggy nervous, but this time she was sure that something dreadful had happened and, worse, that she was the cause. She turned in a panic to Mary, who was calmly straightening out a kink in one of the lines. "Could I set these bells off by anything I said? Is he responding to something I said?" she asked, and Mary responded, "Dear, you don't have that power."

Peggy gave a rueful laugh at the memory. "I've thought of that so much since. It was like absolution in a way. So that really stuck with me. I said to myself, even when he went out on his own: 'Remember, Peggy, you do not have that power.' And it's the truth. You do not have that power."

CHAPTER NINE

The Scourge and
the Silver Standard

Nancy Parent was exhausted. She was commuting between Tom's hospital room and their home in western Massachusetts, trying to be there for Tom and also for their daughter, Ashley. She would stay with Tom until nine or ten at night, then drive home and be up in time to give Ashley breakfast and send her off to school before heading back to Boston. The only time she had to herself was during the long drives across the state, though it was probably for the best that she had plenty to occupy her mind. Otherwise she would have drowned in the guilt she felt that what had happened was all her fault.

She had spent the entire day Sunday, Tom's first full day in the hospital, at his bedside, helping him eat, helping him get as comfortable as possible in bed. Though comfort was far from possible. The sheets were soggy with silver nitrate solution, which drenched the dressings that extended from wrist to elbow on both arms and wrapped around his waist. It was hard for him to shift his weight in bed since the least bit of pressure on his "love handles" was excruciating. The worst, though, was the blister on his left hand.

It was the size of a softball, Tom remembers. Right there in the palm of his hand, as though he had caught it straight off a Sunday slugger's bat. Everyone was amazed—the nurses, the residents, even the attendings. One of them—Tom thinks it was John Schulz—said it was the biggest intact blister he had ever seen. And they wanted to keep it intact as long as possible. Although there are differences of opinion on what to do about blisters, just as there are on most topics related to burn treatment, the practice at MGH is not to break blisters unless they span joints and are likely to burst when the joint is moved. Nor are they collapsed through aspiration with a needle. Both are considered invitations to infection.

A blister is a fluid-filled pocket that forms between the epidermis and dermis. The fluid is the plasma that has escaped through the capillary walls, and as everyone has known since Oliver Cope, it is sterile. The blister acts as a protective pillow, albeit a fragile one, for the raw and angry wound beneath it. Once it bursts, the gate is open for infectious microorganisms. It also hurts like hell. The flap of dead skin has to be removed, the wound cleaned and then bandaged to protect it from contact, and especially contact with air, which is both painful and a source of bacteria.

That is why it made both Nancy and Tom nervous when the occupational therapist came in and started the hand exercises. You might think they would wait for a couple of days at least, but Tom and Nancy found out that rehabilitation therapy starts pretty much from day one. The occupational therapist makes the splints and positions limbs as well as working on small muscles and manipulative function—activities of daily living, or ADLs. Everyone wanted to be sure that Tom would have the use of all his fingers and full strength in his hands once his burns healed. There was a real possibility that the skin would contract into tough, ropelike scars, deforming his hands into stiffened claws. OTs use a combination of splinting devices, massage and manipulation, and exercises to keep the healing skin supple and the muscles stronger than the contracting skin. The goal is fine motor control that is as precise and deliberate as it was before the injury. The degree of ingenuity that goes into their work is extraordinary. OTs play Rube Goldberg to the physical therapist's drill sergeant.

Physical therapists work on large muscles, ambulation, restoring strength and functionality to limbs that have been immobilized, the skin and often the muscle stripped, first by the flames, then by surgery, and possibly by catabolism. Every day, for easily an hour at a time, a PT will work with each patient on Bigelow 13, conscious or not, in pain or not, wrapped from head to toe or on the way out the door. Pushing, turning, working on strength and range of motion, doing whatever is possible to retain what functionality there is and to slowly begin to restore what was lost. For the PT, it is hard physical labor, especially when it involves working on a patient in the steamy confines of a BCNU. For the patient, it is pure agony.

It is no surprise that the turnover among PTs and OTs is greater than among other members of the burn team. The two therapists I watched at work have already moved on, one to spend more time with her young children, the other to get an MBA in health care administration, which he hopes to use as a springboard to a consulting career. It seems a shame to lose two extremely talented patient care professionals, but this is a job for the young and the strong. Like athletes or dancers, physical therapists have to prepare for a time when their bodies are no longer up to the task. "It's exhausting," Mike Tiffany told me a week before he left to start on his MBA. "Up to a few months ago, we had therapists that were in their sixties, but it was much harder for them, especially working on a unit like this where you're lifting 200-, 300-pound people for half of the day. I can't see doing it forever."

Part of it may be the emotional toll. Rehabilitation therapists cause a lot of pain; hour after hour they do battle with recalcitrant limbs and the people they are attached to. "Most of the patients don't like me—at least while they're here," Mike said. He would follow up with them, sometimes for years, when they came back for clinic. "They come to appreciate me once they're out of the hospital, but they tell me how much they hated me when they were here." More than one burn patient has referred to physical therapists as physical terrorists. "Sometimes you just have to be pretty brutal and say, 'If you don't do this, you're never going to walk again.'"

The truth is, it is a lot easier to work with the patients in drug-induced comas. They cannot cooperate, but they also cannot fight. All the therapist has to contend with is the heat of the BCNU or the OR, where he will often do a session after the patient is completely knocked out and everyone else is preparing the room for surgery.

Tom Parent did not have to be told how necessary the five-finger exercise was. In the course of his work as a news cameraman, he had run into people who had been burned, he had seen hand burns that healed badly and he dreaded having the same thing happen to him. "Your fingers and everything else, they contract and they just become nubs. They get webbed and they have to do surgery to make them fingers again." So when the OT asked Tom to move his fingers—his fingertips first, then down to the second knuckle—he did exactly what she told him to, even though it hurt, even though it meant pressing down awfully close to that blister.

Nancy had spent the first night in the motel the social worker had arranged for her, but she knew she had to get home Sunday night. Ashley was supposed to march in the Memorial Day parade the next day. She was in the school band, and she had been practicing and looking forward to this for months. Nancy wanted her daughter to do something normal, and she wanted to be standing on the sidelines watching, just like a normal mother.

So she made the hour and a half drive to Ludlow and picked up Ashley from her brother's home. That night, Ashley said to her, "Look, there's something that everybody knows except me. I don't know what it is. I don't know if Daddy's going to die."

Tom had called her from Cape Cod Hospital, just so she could hear his voice. "I'm fine. They've got me under control," he told her as Nancy looked on, amazed that his voice never wavered despite the pain that was shaking his body. "They want to take me to Boston. I'll probably be there a few days. But I'm fine." But now it was a day later and Nancy knew it would be more than a few days before Tom was home. She was relieved Ashley had spoken up. It was true, they were keeping something from her: the fact that Tom had actually caught on fire. Nancy was convinced that the image of Tom as a human torch would be too horrible for Ashley to bear, but now she was asking and it was time to answer her straight. "So I said, 'I'm not lying to you. Daddy's fine. And I'll tell you what it is.' I told her and we sat there and cried for a really long time. But after that she was okay." Okay enough to drive back into Boston with Nancy on Tuesday morning, to actually walk onto the burn floor and into Tom's room and see what he looked like, how he was, what was going on.

Ashley was relieved to see Tom sitting up in bed, joking with the housekeeper, chatting with his roommate's visitor, being his usual friendly self. And she was relieved to see his face light up when she tiptoed into the room. When the staff found out Ashley wanted to be a doctor, they asked her if she wanted to stay and watch the dressing changes. That was too much. Ashley was okay seeing her father covered with bandages, stained brown though they were, but the site of his raw flesh, well, "It ended her medical career," Tom said.

Still, Ashley came as often as she could, as often as Nancy would let her, considering that she was supposed to be going to school. Nancy would bring her for two days and an overnight in Boston and then drive her back to Ludlow and return by herself for the next few days. She did that for the first ten days Tom was on Bigelow 13, then she went back to work mornings and made the drive round-trip so she could spend the afternoon and evening with Tom.

Neither Nancy nor Ashley was there when Tom's blister burst. He was sitting in the chair watching television, his left arm bent at the elbow. All of a sudden he felt something warm and moist and he looked at his hand. Some kind of liquid was draining out of a tear at the bottom of the blister, collapsing it like a day-old balloon. It was bound to happen eventually, but, "Oh my god, that was . . ." Tom remembers the pain but cannot summon the words to describe it. "They had to come in with the scissors and the knives and stuff and they had to cut . . . that was a tough thing."

Tough, but absolutely necessary.

* * *

If you were asked to invent the perfect growth medium for bacteria, you would be hard-pressed to come up with something better than a burn wound. Put that wound in a hospital bed and add some IV lines and you have won yourself the blue ribbon. A burn wound is essentially a puddle of protein and dead cells, a stagnant backwater devoid of immune defenses and bypassed by the river of oxygenated blood that is supposed to cleanse the area and carry away the debris. Simply put, it is a microbe's paradise.

The planet is crawling with infectious bacteria, and so is the skin of any normal healthy human being. Bacteria are particularly abundant in some environments, hospitals among them. We are not talking about poorly equipped hospitals in developing countries, either. The most modern, up-to-date hospital in the most medically and technologically advanced country in the world is, for all intents and purposes, a giant petri dish. Between what the patients and everyone who works or visits there brings in from the outside and the massive amounts of disinfectants and antibiotics used within the walls, bacteria, and notably resistant bacteria, have taken up permanent residence. It is not such a big deal for most people, who have intact skin and strong immune systems, but for burn patients it is a time bomb with a fuse that is just about a week long.

Not only has their protective skin layer been breached, but their local immune defenses have been burned to a crisp and their systemic immune resources decimated.

The first few days after a major burn are chaotic. That is when patients are at risk for shock and are developing edema, when fluid resuscitation and monitoring are in high gear, when wounds are evolving, when patients with massive burns and insurmountable respiratory damage die. But for those who survive, it is the relative quiet before the storm of sepsis strikes and they start to get really sick.

The surface of a burn wound is sterile at first, all microorganisms fried along with the skin. Over the course of the next day or two, it begins to be populated with pathogens. The first settlers are gram-positive bacteria, the common *Streptococcus*, *Staphylococcus*, and *Clostidrium*, followed after about a week by gram-negative bacteria: *Pseudomonas*, *Klebsiella*, and *Escherichia*. Some bacteria remain on or near the surface of the wound, but others secrete tissue-destroying enzymes that clear a passageway deep into the wound so they can gain access to the circulatory system and rapidly spread through the blood. Thus does a local infection turn into sepsis. Finally come fungi and yeasts, *Candida* and *Aspergillus* being the most common. These opportunists thrive in an environment where immune defenses are meager and beneficial bacteria outnumbered by pathogens.

Even with a relatively small burn and the best of care, Tom managed to brew up an infection within a few days of his arrival on Bigelow 13. Nancy thinks he probably got infected when he put his arms underwater in the men's room of the police station back on Cape Cod. The cool water felt good on his scorched flesh, "but there are just so many germs in the water and there was just no skin there." Whether or not that was the source of Tom's infection, there was no question he had one. "I walked in one day and there was black grease pencil all over his arm. It had started to swell and was very red." The red began on his upper arm, in the vicinity of his armpit, and was advancing down toward his elbow and beyond. "They were tracking by date, by the hour how this thing was moving, and they didn't want it to go downward toward the burn, though eventually that's what happened. So he had a terrible infection." They put Tom on powerful antibiotics. "Vancomycin, straight up," he said in his characteristic clownish manner. "They said it was the third strongest antibiotic in the world," Nancy added. "And I said, 'Why aren't you using the strongest?' After a few days he still wasn't getting any better. He was very sick, and getting sicker."

It was around that time that Nancy got a call from her brother with a message from a friend of his who had been badly burned as a teenager. "He made my brother call me, and my brother said, 'I didn't want to tell you this, but he insists that I call and tell you, you have to be prepared that week two is worse than week one, and week three is worse than week two.' That's something that's unique with burns. You expect when you go into the hospital and treatment begins that things are going to keep getting better. I would say to Tom every day, 'We're one day closer, it's one more day and we're that much better,' and then we got into week two and we realized we weren't. I think week two was kind of a bleaker week for us because we felt like we were over the hump, and then realized we weren't over the hump. We were in probably the worst part of it."

It has ever been thus. As long ago as 1833, the noted Scots physician Sir George Ballingall observed that victims commonly died three to six weeks after being burned, "sinking into a hectic state, exhausted by a profuse discharge of matter from an extensive suppurating surface." That was, of course, long before the age of antibiotics, even before James Lister introduced the concept of antisepsis in 1867, so there was little to do but nurse burn patients as they made their way toward death. Preventing burn infection continued to be a challenge well into the twentieth century. From the complicated and misguided use of triple dyes to the simpler and only slightly more effective gauze wrapping promoted by Harvey Stuart Allen and Sumner Koch, nothing the best minds of that generation could come up with was equal to the ever present threat of invasive bacteria. Even in the postwar period, when penicillin and sulfa drugs began to tip the balance in the fight against infection, bacterial sepsis claimed a disproportionate number of lives. In the decade between 1966 and 1975, it was the cause of death in as many as 80 percent of patients with burns larger than 50 percent TBSA.

John Moncrief, who was director of the U.S. Army Institute of Surgical Research in the 1960s and president of the American Burn Association, vividly described the demoralizing state of affairs. "Prior to effective topical therapy, the burn wards were known for their repelling odor, the fatiguing duty, disconcerting frequency of fatal septic complications, and the often pungent smell of frequently used disinfectants. The patient population that survived the prohibitively high mortality rates were drawn and tired, wasted, hollow cheeked and glassy-eyed, disinterested in surroundings, often disoriented, anorexic, and even under the best nursing conditions frequently inundated with gross infection and odor."

From time immemorial, infection has been a scourge. It remains the cause of death of many burn patients, even with today's powerful antibiotics and medicine's deep understanding of the natural history of burn infection. The bigger the burn, the greater the risk of infection and the more serious that infection will be. As with so many other complications of burn injury, the very young and the very old are the most likely to develop life-threatening infections.

It is not just infections in the wound, however. Bacteria and other pathogens can invade the body through the wound or through another point of entry, or they may be there already. The lungs, especially but not only in patients suffering inhalation injury, are highly vulnerable to infection. Merely lying still in a hospital bed for any length of time is a major risk factor for pneumonia. One of the most insidious sources of bacterial infection following a burn is the gastrointestinal tract, which is normally colonized with bacteria that aid digestion and the synthesis of vitamin K, which plays an important role in blood clotting. The GI tract is a closed system, albeit a long one, with one entryway and one outlet and miles of tubing in between. Something similar to the hyperpermeability that permits fluid to escape from the blood vessels affects the walls of the intestines, making them permeable to their resident bacteria. Although harmless and beneficial in the gut, these bacteria are dangerous when they wander elsewhere in the body.

Necessary interventions by medical personnel are another source of bacterial contamination. Anything that goes through the skin pierces the microbial barrier, IV lines being a classic example. The fine plastic catheters used for lines act as toll-free tunnels into an immune-compromised body. Bacteria collect around the entry point, colonizing and contaminating the surrounding tissue. That is why lines are changed regularly and their location is chosen with care. The preference is always for an unburned site, since a burn or any other open wound is a breeding ground for infection. As John Schulz explained it, the best place for a central venous line is the internal jugular vein, located in the neck. Second best is the subclavian vein, which runs below the collarbone. Last choice is the femoral vein, accessed through the groin, which is second only to a wound in terms of the potential for infection. Schulz said he likes to avoid the femoral vein if at all possible, but "burn trumps all: Go to the unburned site first because it is least likely to get infected."

Tom Parent did not need central and arterial lines, but he did have an intravenous line to pump in fluid and pain medication. IV lines are usually peripheral—in an arm or a leg—but priority still goes to unburned skin. It is not always possible to find unburned and intact skin, however; some patients do not have much of it, and after a while, what little there is has already been used. Tom's arms, for example, were clearly off-limits, so that left his legs. As Tom remembers it, after a few line changes, they started to run out of places to plug him in.

* * *

The same thing that makes a burn wound a fertile breeding ground for infection makes it useless to give antibiotics by mouth or as a subcutaneous injection or an intravenous drip, unless bacteria have actually penetrated the wound or spread systemically. The blood that would carry the drug to the wound simply is not getting close enough to where the trouble is to have any effect. That is why antibacterial substances have to be applied topically, to the surface of the wound.

This is not a new idea. As John Burke put it, "You put some slop on the burn wound and hope that will keep it from getting infected. This had been going on a very long time, since Adam probably. It seemed to be a reflex of people that if you have something that's wrong with you to put something on it. Whether it makes any sense or not is neither here nor there. At any rate, a lot of things were put on—tannic acid, all kinds of poisons, one thing or another."

In 1799, James Earle observed that "there are few accidents or maladies to which mankind are subject, which have met with a greater variety of treatments than burns." A century and a half earlier, a herbal handbook, the *Cruydtboek*, by Rembertus Dodonaeus, listed seventy-three preparations for treating burns. The catalogue has only grown longer in the intervening years, with new products coming on the market every day. Every pharmaceutical company on the planet has a new improved cream or ointment or liquid or dressing or temporary skin substitute. Enormous sums of money are spent on convincing the holders of the purse strings and the writers of the orders to give them a try. After the Rhode Island fire, Mary-Liz Bilodeau's phone was ringing off the hook with calls from salespeople. "There's always new dressing materials coming out. I was amazed at the number of people that were calling: 'Try this new product, try that new product, we'll give it to you for free.' Because all they want is to be able to say they provided dressings for the patients from the Rhode Island fire. And I said, 'No, this is not the time to introduce a new product. I don't have time to babysit it. I'm going to use things I know work.'"

Mary-Liz is not hidebound about wound care. She probably knows as much as anyone in the burn care community about when to use what on what kind of burn, but she says, "There's so many ways to do burn care; there's not necessarily one right or wrong way to do it." Burn centers tend to use the materials and methods they are most familiar with, the ones that have worked for them. Economics plays no small part.

The way they do burn care at the General relies, remarkably enough, on silver, a precious metal people have been using for medicinal purposes, if not since Adam, at least since classical times. Ensuring a safe drinking water supply in ancient Rome was a matter of using silver vessels for storage or tossing a few silver coins into the water barrel. Back then, everyone knew it was an effective disinfectant, though they did not know why. Medieval physicians observed that wounds were less red and healed faster when silver was applied. The apocryphal writings of the possibly fictional monk and alchemist Basilius Valentinus mention the healing powers of silver. It is suspected by some that Valentinus was the creation of Paracelsus, the Renaissance man of medicine, who had something of a talent for self-creation. Born Aureolus Philippus Theophrastus Bombastus von Hohenheim, he was widely known as Paracelsus, not so much in tribute to Celsus, the Roman physician of the first century A.D., but to assert that his skills surpassed those of his predecessor. He disavowed medical practices based largely on superstition that had survived the fall of the Roman Empire and persisted throughout the Middle Ages. In his 1535 treatise on surgery he observed, "If you prevent infection, Nature will heal the wound by itself." Whether or not Paracelsus and Valentinus were one and the same man, they both had silver in their medicine chests.

Nitrate of silver is included in the shipboard kit of *The Surgions Mate*, a text for ship surgeons published in 1617 by John Woodall. At sea or on dry land, it was widely used as a topical treatment for chronic wounds. Usually, it was dried into a powder, formed into sticks and, like styptic pencils, dipped in water and dabbed on wounds. The harsh nature of this highly concentrated form is suggested in the range of names by which it was known: lunar caustic, *lapis infernalis, pierre infernale,* and *Höllen-stein*—hell stone.

Silver was used in this manner to enhance wound healing throughout Europe well into the nineteenth century. It was only in the 1870s, however, that its antibacterial properties were recognized. (The concept of a bactericide never entered anyone's mind until Louis Pasteur and Robert Koch identified bacteria as agents of disease.) By 1881, however, silver was proposed as a treatment for gonorrhea, an application that survives in the practice of instilling a weak solution of silver nitrate in the eyes of newborns to prevent blinding infections that can be acquired during the passage through the birth canal of a mother infected with *Neisseria gonorrheae* and the even more common *Chlamydia trachomatis.*

Karl Credé, the German obstetrician who introduced this prophylactic use of silver, also used it in novel ways for dressing wounds. He devised what he called a "white silver dressing," a piece of gauze faced with silver foil and applied foil side down. His "gray silver dressing" was made from sterile cotton liberally sprinkled with powdered silver. By 1895, the noted American surgeon William Halstead had brought silver to Johns Hopkins, where he advocated wrapping wounds with silver foil and using silver sutures in surgery. Silver foil dressings were used throughout World War II after which the precious metal was eclipsed by the new miracle antibiotic drugs. Only in the treatment of burns did silver continue to be used. Silver nitrate in what we now know was a dangerously strong 10 percent solution was applied to burns, with and without tannic acid, in the years leading up to the Cocoanut Grove fire.

There is plenty of science to back up the use of silver as an antimicrobial agent, though no one is certain how and why it works so well. That is not as unusual as it may seem; many drugs, including aspirin, are known to be effective even though the mechanism of action is not fully understood, if it is understood at all. In the case of silver, we do know that it is biologically active only in its ionic form. That is, as an electrically charged particle, such as occurs when it is in solution or bound to a nitrogen salt as it is in silver nitrate. Ions readily attach themselves to molecules of other substances. It appears that the silver ion binds to molecules on cell membranes, interfering with the cell's ability to breathe. Fortunately, its affinity is to prokaryotic (bacterial) cells, not mammalian ones, making it selectively bactericidal. Even more fortunate, it not only kills all bacteria, gram-negative and gram-positive alike, it does so in a way that does not foster the development of resistant strains.

If it sounds like ionic silver has it head and shoulders above even the broadest spectrum antibiotic drugs, it is important to remember it is a surface treatment only. It cannot penetrate eschar and it certainly cannot be used internally. Silver is not toxic externally or internally; when combined with nitric acid, however, the resulting compound is caustic to the skin, especially in high concentrations, and toxic if swallowed.

The 1960s marked another watershed moment in burn care as new topical agents began to turn the tide against infection. The first effective modern topical antimicrobial was a sulfa drug with origins stretching back to the darkest days of the World War II. German military doctors used a drug known as marfanil to treat gas gangrene, a common but deadly battlefield infection. The formula was part of the booty captured by the victorious American army, which handed it over to a team of researchers at the U.S. Army Institute of Surgical Research. They changed the name to mafenide and studied its effectiveness against the most common bacteria found in wounds.

The good news about mafenide, which was marketed in the mid-1960s as Sulfamylon, is that it penetrates eschar and is thus effective against invasive bacteria. That is balanced by several bits of bad news. The drug is rapidly absorbed, requiring frequent reapplication, which is painful wherever nerves have survived. When absorbed, it may cause adverse systemic effects as well as allergic and other severe reactions in people sensitive to sulfa drugs. Although it is effective against a broad spectrum of bacteria, many develop resistance to it, unlike silver compounds, and it does not combat fungal infections at all. Still, at the time it was the best agent available for what continued to be a scourge.

At about the same time, Carl Moyer and his colleagues at Washington University Medical School in St. Louis were taking a new look at silver nitrate. They experimented with a far more dilute solution than had been previously used and found that a 0.5 percent solution was strong enough to kill bacteria and fungi in burn wounds without damaging healthy tissue or interfering with healing. On the downside, it leaches electrolytes from the open wound, requiring constant monitoring and frequent replacement of these vital blood salts.

Clearly, there was a need for an agent that combined the virtues of the two topicals without the vices of either. Half the distance was traveled when silver sulfadiazine was developed in 1968 by Charles Fox at Columbia University College of Physicians and Surgeons in New York. Silver sulfadiazine, marketed as Silvadene but available as a generic now that its patent has expired, is a water-soluble cream that contains both silver nitrate and sodium sulfadiazine. Unlike silver nitrate solution, it does not stain, and unlike mafenide, it is not painful to apply. It is effective against a broad spectrum of bacteria, but some bacteria develop resistance to it. It remains active for as long as twenty-four hours, so it can be applied less frequently than its counterparts, but must be removed completely before reapplication, a process that can be very painful since it does not easily rinse off. It has to be scrubbed off with sterile cloths or gauze sponges or with forceful jets of a warm water. It has fewer side effects than mafenide, but because it contains sulfur compounds, cannot be used by some patients. Although it is less than perfect, silver sulfadiazine is now the most widely used topical antimicrobial for burns.

More recently, a high-tech stepchild of the silver foil and silver powder–impregnated dressings of Karl Credé promises a completely safe, easy-to-use, nonstaining, and long-lasting way of delivering silver to the surface of a wound. Acticoat is basically a gauze pad coated with silver dust, though the gauze is a polyethylene mesh bonded to a thin plastic film, and the silver particles are dubbed nanocrystals. Because they are very tiny, their combined surface area is very large, the better to latch onto bacteria in the wound. So long as the dressing is moist, the tiny crystals act as thousands of sustained-release silver ion bullets. The release lasts for several days, so theoretically daily dressing changes are not necessary, but if an infection develops, two or three days are enough for it to get a firm foothold. The stuff is very expensive, though it is cost-effective in the long run since dressings are changed less frequently. Although not practical for big burns, it is very useful for smaller wounds, especially in outpatient settings.

At the present time, silver sulfadiazine is used by the truckload in just about every burn center except Mass General. Most consider silver nitrate old-fashioned, messy, and difficult to use. The fact that it turns everything it touches a brownish black is a major turnoff. Silver nitrate dressings must be kept wet, which means checking them every two hours and dousing them with more of the solution when they start to dry out. High heat and humidity keep evaporation and heat loss at a minimum. That works if the patient happens to be in a BCNU, but it is more of a problem elsewhere, whether it is out on the floor on Bigelow 13 or in another burn unit.

Although silver nitrate is the mainstay on Bigelow 13, silver sulfadiazine is used when the situation warrants. Outpatients coming to the clinic who have small burns are instructed in how to apply the cream and how to remove it. Similarly, sulfadiazine is prescribed for patients ready for discharge who still have small areas that need topical treatment, and even the occasional inpatient. Although it is expensive, Acticoat is used as well. Mafenide is used infrequently and only when bacteria have invaded the wound.

Silver nitrate goes hand in glove with the BCNUs and the entire design, operation, and philosophy of burn treatment at the General. The cornerstone is the practice known as early excision, the prompt surgical removal of eschar and all dead or dying tissue. All that remains is a wound bed on which new skin will either regenerate or be grafted. Early excision is hardly unique to the General—it is the standard of care in burn treatment—but the use of silver nitrate in conjunction with it is. No one can explain it better than John Burke, the mastermind behind the burn center.

"The silver nitrate method fits so comfortably into the prompt excision and immediate wound closure system, and that is the thing that is the most important. That really is what saves people's lives and gets them well quicker and gets a better functional and cosmetic result," Burke explained.

"There's a great deal of misunderstanding about why we use silver nitrate. The simple truth of the silver nitrate business is that it's not because we feel silver nitrate is the world's best topical material. It isn't that we don't realize that silver nitrate doesn't penetrate an eschar and if the bacteria get below the surface then silver nitrate's not effective. Why we use silver nitrate is that burn care is very complicated, involves a great number of people in the burn team, and all of the parts of the system have to fit together, so that if you're going to promptly excise a burn and close the wound immediately after you've excised it, then you find that it is an enormous difficulty putting an antibacterial cream on that. But if you put silver nitrate on the dressing, that fits right in with the system. And since we don't intend to have eschar sitting around, penetration doesn't really ring our bell. Silver nitrate is just part of the system that allows you to do the primary excision."

* * *

On the face of it, Edna Gavin and Carl Baxter could not be more different. She is petite, dark, and elegant. He is burly, blond, and bearlike. She speaks with the lilting sounds of her native Jamaica. His voice carries echoes of the peace, love, and whatever of the 1960s. She has a laser-sharp grasp of what a wound needs and why. He takes a Zen approach to his work, his technical expertise flowing from a more intuitive sense of his patients' emotional needs. What they have in common is a deep commitment to the patients they minister to and the wounds they help heal.

They are the two senior patient care associates on Bigelow 13—the burn techs who do most of the hands-on care for the floor patients, helping them with meals, bathing, and personal care, checking temperature, blood pressure, and so much more. PCAs were once known as nurse's aides, though the work Edna, Carl, and the other burn techs do far surpasses the usual definition of the term. For one thing, they spend enormous amounts of time with patients who are conscious. These patients may be in pain, they may be scared, but they are awake and the techs are the ones who help them get through the difficult days. There is no question that the techs are pivotal members of the burn team.

"Our burn techs are *wonderful* with wounds," Nancy Giese told me. "Edna has wound care down to a science and beyond. She knows how to evaluate and to head things off before they become a problem. And we rely upon her to tell us." The work Nancy and the other burn nurses do simply would not be possible without the burn techs. "I'm here in the unit, taking care of this guy and this guy's wounds, and I can't physically see every inch of every wound on the patient that she's got out there.

Often I'll go out and do half of the dressing changes with her because I need to see what the patient's wounds look like because I am the one who has to document on it. But I'm not poring over it for two hours like she is. So we have to have a working relationship and a trusting relationship. I know she knows what she's looking at and how to tell me what she sees, and then we can address it together."

Edna and Carl have been working at the General for more than thirty years combined. Edna worked in burns when it was on White 12; Carl came to Bigelow 13 a dozen years ago. Other techs come and go, the ranks swelling and shrinking according to whatever calculus Tony Di-Giovine is grappling with, but it is these two everyone depends on. Even John Schulz defers to Edna when it comes to wound care. Edna says she has an excellent working relationship with all the attendings, but if one of them suggests a dressing she does not think is best, she will make a face or just out-and-out question it. "Dr. Schulz will say, 'Edna, what do you think?' And I'll say, 'How about a little nitrate for a couple of days,' and he'll say, 'All right, Edna, whatever you say.'" And that will be that.

It was Carl Baxter who took care of Tom most of the time, did his dressing changes, and taught Nancy how to do them once she took Tom home. "Carl was just unbelievable. Nobody could change a bandage like Carl," Tom told me. "He was fast. He was good. He got it done. He was in, he was out, and he knew what he was doing. He used to say to me, 'You know, I just want to keep the pain to a minimum.'" But more than that, "Somehow he could sense when I was getting down or whatever, and he would come in and just talk." He also talked to Ashley, helped her deal with missing her dad and worrying about him.

Carl considers that part of his job. "It's very difficult for the family, the loved ones, the patient. Everyone is in shock, traumatized by the sight of the wound. So anything that you can give to them that will help them cope and get through their adjustment and for me to get the job done, and maintain as a human response back and forth." Listening to Carl talk about his work can be like unraveling a Zen *koan*, but there is no arguing with the bond he made with the entire Parent family. "Until the day I die," Nancy said, "he can call me at any time, if he needs something, and I'll be there."

Tom's dressings were changed twice a day, at 10:00 A.M. and again twelve hours later. The morning change was the worst. That is when the docs came around to look at his wounds, which meant taking down the dressings in advance and waiting, his raw flesh exposed to the air, until the inspection team arrived. Even though he would get extra pain medication twenty or thirty minutes ahead of time so it would have kicked in by the time Carl started peeling back the gauze, Tom still dreams about the pain. One of the doctors once explained to Tom and Nancy that pain meant healthy nerve endings, that the damaged ones were repairing themselves, that new ones were growing. "It got to the point that it was just a bad joke," Tom remembered. "I know, I know, it's good that I have pain," but he also knew he just had to suck it up. And wait for the doctors to arrive.

One time they were taking longer than usual to get around to room 42, and Nancy Giese—he called her Nurse Nancy or Tall Nancy, to distinguish her from My Nancy, his wife—went down the hall and told them to hurry it up because her patient was waiting and he was in pain. "Because I'm in a sterile environment—you know, gowns and masks and everything before you could come in—they would have to drape a sterile cloth over my arm until the doctors came. So I figured out a way where my hand was up on a pillow and I would take my thumb and would keep it up here and they would make a little tent because if anything touched my arm, it was excruciating. I mean, that's the only word left to describe it."

If waiting for morning inspection was bad, debridement was even worse. Every time his burns were exposed, the sticky ooze had to be wiped away and all dead tissue removed. The ooze—technically it is called exudate—is a porridge of protein and dead cellular matter, the muck of the battlefield that was his wound. Combined with eschar and shreds of dead skin gradually separating from the wound bed, it forms an ideal culture medium for bacteria. None of this once living matter will ever recover. It serves no useful purpose and can only do damage, so it really has got to go. The way to get rid of it is with debridement *(dee-breed-ment)*.

Debridement can be done in a variety of ways, depending on the depth, extent, and location of the burn, and the expectations for healing. The goal is to remove all devitalized tissue, exposing healthy tissue that will either heal on its own or become the foundation for grafted skin. In earlier times, maggots were employed to eat dead tissue. These days it is done by mechanical, chemical, or surgical means, or some combination of the three.

At just about every burn center in the United States, the lion's share of mechanical debridement takes place in the tank room, an echoing space equipped with stainless steel showers, tubs, and tilt tables. All patients—be they ambulatory or in critical condition—go there at least once a day to have a solution of warm water and antibacterial wash sprayed over burned areas, loosening crusted exudate, dead surface tissue, and the previous day's topical antimicrobial cream. Nurses and technicians scrub the wounds with sterile gauze sponges to finish the job. It is an agonizing process, even when pain medication is stoked up in advance. It can also be bloody as debridement gets nearer to healthy tissue and newly formed capillaries. A burn survivor still remembers her tank room sessions, more than twenty years later: "It felt like a wire brush raking against my raw flesh. I would scream and cry and beg them to stop. I can still hear my own screams today." In that regard, the critical patients, the ones who are unconscious, have it better because they have no idea what is happening.

There is no tank room on Bigelow 13. John Burke banned the practice in 1974 when burns were still treated on White 12. He considers it a sure-fire recipe for infection. "Most people try to solve the cleaning problem by putting somebody in a tub. Well, the literature is full of very well done, absolutely perfectly controlled studies that say that the worst thing you can do to a patient is put him in a tub. If you really want to get the bacteria nice and evenly over everywhere in the burn, put him in a tub." So at the General, wounds are cleaned and debrided at the bedside. It is still painful and sometimes bloody. It takes a long time and has to be done over and over again. But the nurses and techs would not have it any other way.

Get anyone on Bigelow 13 started on the subject of tanking, and you will get an earful. The gist is that it is barbaric, torture to nurse and patient alike. Moreover, they say, putting burn patients in a tub of water is tantamount to dunking them in a bacterial soup. It does not improve matters much to have them stand or sit in a shower or lie on a stainless steel tilt table while they are hosed down. Listen to Bob Droste, who arrived at the General during the waning days of tanking, or tubbing as it is sometimes called. "I think it's a terrible idea. Before Dr. Burke, we had a tank. It was a big stainless steel box, and we put in a plastic liner. Each patient got a separate liner, so there was no contact between the patient and the tank. Now assuming that there's no cross contamination between patients, you still get autocontamination. If there's bacteria here," he pointed to his chest, "it's now everywhere."

"The tanking that they do now is not actually a tank. You hose from head to toe. So if it's here," he pointed to his chest again, "it's not going to end up on the head maybe; it's going to end up going downhill. But if it's here," pointing to his forehead, "it's everywhere. So there's still a lot of autocontamination. And," he lowered his voice ominously, "it's brutal. You have to move the patient from their room to this thing. They're open and exposed. It's very hard to keep them warm. And it's very hard to do things in an aseptic fashion." Bob conceded that he is most comfortable with the way he has been doing it for years. He is the first to acknowledge that there are many different ways to do wound care, and no one has ever proven that one is always better than the others. The best way is the way the staff is accustomed to and the facility is set up for. "Our way works, but it's way too messy for a lot of places. My bias is that I've never worked anywhere else, except for a short couple of periods. All I know is this, so this is right, according to what I know."

On the other hand, the nurse manager of one of the many burn centers that do tanking told me the patients love their daily bath. It is referred to as hydrotherapy, which makes it sound less like medieval torture. Pain control is much more aggressive today than it was twenty years ago, and everyone is sedated in advance, so pain is minimal. Furthermore, all tank room equipment is thoroughly disinfected and an antibacterial agent is added to the water. It is hard to argue with a practice that is followed in the overwhelming majority of burn treatment centers in the United States and throughout the world. In the absence of a clinical study comparing infection rates in burn centers that do or do not do tanking, it is a prime example of there being more than one way to treat a burn.

Aside from scrubbing with gauze and jets of water, mechanical debridement happens automatically when dressings are removed. Anyone who has pulled an adhesive bandage off an open cut can attest to that. When Carl Baxter takes down a dressing, he tries to gauge how much each individual patient can tolerate. He will start by pulling, "but if the person is in agony, I'll soak it for a while so it's easier to take off. The more we can remove the tissue that isn't living and get the medicine on the wound itself, the quicker it heals." He picks and pulls and scrubs, keeping up a conversation as he works. "I say: 'We will get through this.' I never say 'you.' I always say 'we.' They're suffering. It's very sad. And you have to work with that and stay hopeful and try to project that to the patient as much as you can." He may work for an hour before wrapping up the wounds again, and then tell the next shift how much he accomplished so they can pick it up from there. Debridement is a continuing project; every dressing change is another opportunity to remove more debris.

The body itself helps by secreting chemicals that digest dead cells. When this happens it is called autolytic debridement, and it takes place when a wound is protected by an occlusive dressing, which keeps moisture in and bacteria out. Autolytic debridement, which is neither painful nor harmful to healthy tissue, relies on proteases, which in this instance are doing a service by selectively liquefying nonviable tissue. It is a slow process, however, that must be monitored closely since the same conditions that foster autolysis are hospitable to infection.

Enzymes from plant, animal, and even bacterial sources are also used for chemical debridement. Ointments containing papain (a proteolytic enzyme found in papayas and used in meat tenderizers), collagenase (derived from a bacterial source), or a handful of other enzymes digest dead tissue, making it easier to wash away. Like autolytic debridement, these enzymes require a moist environment, so the wound must be covered with an occlusive dressing after the ointment is applied. Enzymatic debridement is faster but less selective than autolysis; some enzymes damage healthy tissue at the margins of the burn wound, so the ointment must be applied with care. Others are deactivated by metals, including silver, making them a bad fit when wounds are treated with silver nitrate or silver sulfadiazine.

All of these strategies are useful for partial thickness burns, like the ones Tom Parent had, and they are most practical when burns are relatively small. Deep and extensive burns that have developed eschar must be dealt with surgically. Sharp instruments are used to shave and slice away all burned tissue. As with all surgical procedures, this involves blood loss and the need for anesthesia.

Tom Parent had no eschar to be sliced off, but there was plenty of ooze and slough. He developed his own strategy for getting through the twice-daily ordeal, a pain, he says, "that I wouldn't wish on my worst enemy."

He can still remember how the air felt against his forearm when the dressing came off: Like it was on fire all over again, a pain that burned through the pain meds, taking everything he had to keep from crying out. It was as raw as hamburger meat, though it was a lot less oozy than it had been even the day before. Nancy—Nurse Nancy—would dip a square of gauze in the bowl of sterile water. As she lifted Tom's arm, he would slide his foot between the sidebars of his bed, wrapping it around and holding on so tight the front of his foot and toes cramped up. It was the only way he could endure what he knew would come next. No matter how fast she would do it or how gently, that gauze sponge felt like steel wool against what used to be his skin.

"They would take the gauze sponge and they would just scrape, scrape, scrape. After a while, you just know what you're in for, but the way I dealt with it was I would have to think past it. I would say, 'Okay, you know what? In *x* amount of time I'll be past it, and it will be done.' One time I happened to mention this to Nancy. I said, 'How long do you think this will take?' And she said, 'It will be about fifteen seconds.' So I said, 'Okay, I could do that.' So I got all ready, hooked my foot under the bed rail, and she actually counted down the fifteen seconds for me. So that's what I could mentally focus on. I had to disassociate the pain, and I needed to focus on something. I don't know if she kept real good count, but I thought, 'Okay,' and then I could make it and then she would be done. And then I could relax. It's funny, you develop what you need to do to get through it."

* * *

Until the last quarter of the twentieth century most people who survived the first hours after a major burn ended up dying from infection. Here is John Burke's capsule history of modern burn treatment: "When people began to understand what shock was about, and that you could fix the burn shock by fluid resuscitation, the mortality for burns really changed very little, but the time you died changed a great deal. So if you were burned, let's say, 40 percent of the body surface and you were treated by flowers at the bedside, you died on about day three, four maybe. If you were fluid resuscitated—with almost anything—you lasted through the burn shock period and you developed sepsis in the second or third week and you died of sepsis in the third week. Whether living a couple of weeks longer made any difference, I've never been really sure. At any rate, the mortality problem was not for practical purposes changed significantly at all.

"When Sulfamylon began to be put on the wound that didn't change how many people died very much. What happened was it just moved it back a ways so that now you lived for three or four months and then you died of malnutrition and whatever, and finally infection.

"Now it's not that it was a completely useless chase, because even with fluid resuscitation, the people who would be on the borderline of dying would probably not die if their shock was fixed. And the people who would be on the borderline of surviving sepsis would probably survive. So more small burns began to survive, but when you get to the big injuries, with 30, 40 percent of the surface, survival didn't really change very much.

"And it was perfectly clear—this was in the '60s—that burn care was in the dark ages, that everybody who had a significant burn died, and those who didn't die were so miserably and psychologically damaged that their lives weren't worth much even if they did survive."

John Burke was in the vanguard that led burn treatment out of the dark ages. "It was clear that the only way that you could possibly solve this problem was to get rid of the tissue that had been killed by the fire and close the wound with skin. And the only way that you could reasonably, at least in my mind, get rid of all the dead was to cut it off."

CHAPTER TEN

Cutting Off the Dead

It is seven o'clock on a Wednesday morning and the two women are expecting guests within the hour. One of them flips on the boom box in a corner of the room, and they work to the reggae rhythm of Bob Marley's "Buffalo Soldier," looking for all the world like two friends preparing for a party—arranging the furniture, setting the tables, laying in supplies, and keeping up a lighthearted chatter about the sort of things friends talk about: the previous weekend, an upcoming vacation, a pet's antics, a movie star's scandalous divorce. One covers a stainless steel table with a paper cloth, plucked from a seemingly endless supply of sterile drapes, wrappings, curtains, all the same medium blue, that sit in stack upon stack on a shelf in the glass-fronted cupboards that cover two walls of the large room. On another shelf are boxes of blades, packets of sutures, needles, surgical staplers, and staple removers. A bottom shelf holds a case of sterile water and another, blue-paper-wrapped rolls and squares of gauze. There seems to be enough stuff here to last a lifetime, but chances are it will have to be restocked before the end of the day.

Her companion emerges from the anteroom, lined with open shelves like a kitchen pantry, carrying two bundles wrapped in blue paper. Inside are stainless steel bowls, and inside those, wrapped packages of instruments: Goulian knives, Kelly clamps, forceps, enough to serve a party of twelve at least. She wheels a second table into place, making an L-shaped serving area at the side of the operating table. Another blue cloth goes on, then she sets out jugs of sterile saline, and in another bowl, two or three huge syringes that look like turkey basters. Alongside them is a powered dermatome, a contraption that looks and works like a heavy-duty electric cheese slicer, and next to it, a mesher. The analogy to a hand-cranked pasta maker is irresistible. This is the standard equipment for excision and grafting—the knives and blades that will be used to cut off the dead and harvest healthy skin to replace it. The heavy console that powers the Bovie electrocauterizing wand stands against one wall. Later, when it is time to stop bleeding, it will be pushed into place beside the operating table. Metal grounding plates stand ready to be slipped beneath the patient to protect him from being shocked each time the wand zaps a severed capillary.

The music segues to the Grateful Dead as the women keep on truckin' in the supplies—you could not make this up. Charles Jeffrey, the anesthesiologist, comes in and stations himself at the head of the operating table, clipping a blue curtain in place to separate the operating area from his domain. Though there is nothing of the wizard about him—he is tall and slim, low-key but affable—the electronic displays and lengths of tubing, the blips and ticks, the chuffs and gurgles that issue from behind the curtain conjure up the land of Oz. He begins adjusting dials, checking connections, and marking the log he will keep of everything that is done to the patient throughout the procedure.

The job of the anesthesiologist is no more nor less than to keep the patient alive during surgery. Not just unconscious and out of pain, but alive and out of danger. He monitors and logs all vital signs, keeps the patient hydrated, makes sure temperature, heart rate and rhythm, blood pressure, respiration are all within normal bounds. The tourniquet is his territory, so is the respirator. If the patient needs blood, he takes care of the transfusion, and if something goes wrong, he is the one who runs the code. Like John Schulz and the other attendings, Dr. Jeffrey shepherds residents on rotation, imparting as much wisdom as he can during their two-month stay.

This is OR 36, one of the forty or so operating rooms that run through several buildings spanning the third floor of the General. It is the burn room, the assigned territory of the burn service, which uses it regularly on Mondays, Wednesdays, and Fridays. On Tuesdays and Thursdays, it is loaned to other services, but the room is set up for burns, and the nurses, scrub tech, and even the anesthesiologist who work there are all burn experts.

Terri Leddy has been a burn OR nurse for six years, an eternity considering the burnout potential of the job. She spent the previous fourteen years in thoracic surgery. She is a plainspoken woman with a warm manner and a mischievous smile that shows up in her eyes when the rest of her face is covered with a surgical mask. She appears to be completely unflappable, though it was a flap that got her into room 36. "I got pissed off one day at a thoracic surgeon and said, 'I've had it!' There was an opening in burns. A friend of mine who worked there was expecting twins, and she said, 'Just try it.'"

"I never thought I would stay this long, but it's because of the surgeons and the two other people I work with that I'm still there. The worst, most disgusting day in the burn OR can be better than the best day in some other operating room, just for the sheer fact of who you work with. They're a wonderful group of people. They are just so caring, so good to their patients."

Terri is referring to the three attending surgeons—Drs. Ryan, Sheridan, and Schulz—and the two other OR nurses who work the room in tandem, a team of two each day. There is also the scrub tech on permanent assignment to the burn OR, and Charlie Jeffrey, the anesthesiologist, also a permanent fixture, both of them definitely members of the team. And then there are the residents, though to a much lesser extent. They are, after all, there for only two months, averaging twenty-four days in the burn OR, only about half of which coincide with the days Terri is there. Besides, Terri says, "There's not one person that's come through here that I know of that's going to go into burn surgery. Hardly anybody says that that's their goal, hardly anybody is interested in that." Except John Schulz. Terri remembers when she worked in thoracic and he came through as an intern. "He was always just the nicest, nicest guy, very conscientious, as an intern. Extremely smart. He's just a normal good guy. There is not one person on the floor that doesn't like him, nurse-wise. Everyone just loves him." They call him "the number 2 pencil," though not to his face. It is a startlingly apt physical description, tossed off with the typical irreverence of the nurses who work in OR 36. Mike Wilson and Bob Droste remember Schulz as a resident too. In fact, they say, they saw both him and Colleen Ryan come through Bigelow 13 on their way to a life in burns. A lot of silver nitrate has flowed over the dam in the days since.

Terri works two twelve-hour days three weeks out of every month. In the fourth week, she works the desk for five days. Theoretically, those five are eight-hour days, and the task, which she shares with an anesthesiologist, is supervising the entire OR complex. But according to Terri, it is really "6:15 to whenever . . . whenever you're ripping your hair out and screaming: 'Let me out of here!' You're just keeping the OR running as best you can. You do the staffing—you hope your nurses come in—you work with anesthesia to keep the rooms running. It's like a puzzle; you're constantly moving cases around. There's a plan the day before, but usually a third of it changes by the next day, so a lot of it is troubleshooting."

Even the two twelve-hour days can stretch into the evening. It is not easy to get nurses to work the burn OR. "It's a great place to work, but it has such a bad reputation: It's hot, it's gross, the patients smell, and they [staff in other operating rooms] really don't think we do anything. They think it's just taking off skin, but they never come in and actually see what we do do." Nurses who agree to work there are promised no weekends, no holidays, no off-shifts, though they must commit to staying until the case is finished. "If they go to ten o'clock at night, we stay till ten o'clock at night."

Terri recruited the two other nurses she works with, women she considers her best friends, despite the fact that they rarely, if ever, see each other outside the workplace. "We know every intimate detail of each other's life, and would do anything for each other. If it wasn't for them, I wouldn't be there. There's no doubt in my mind. When you know you're going to work with someone you care about, it makes a big difference." After Terri had been in the burn room for about two years, the nurses who were there when she started moved on. So she thought about other nurses on the OR floor she would get along with, and she convinced them to come aboard. Now it is a sisterhood that works like a well-oiled machine, though Terri worries about what will happen when this over-forty crew no longer has the stamina to function for twelve hours in heat and humidity that makes a July day seem like early spring. The nursing shortage is no less acute in the OR than it is at the bedside, with fewer young people interested in the high-stress atmosphere.

For her part, Terri far prefers the OR to bedside nursing. "They're unbelievable, the nurses on that floor. I don't know how they do it. They not only have to do all this intense work with the patients, but they have the family. I could not do it. Bless them is all I can say."

OR nursing is completely different from bedside nursing. One of the big differences is in the personal dynamics. Up on Bigelow 13, the nurses' most intense relationships are with the patients and their families;

for moral and physical support they rely on the other nurses. They pour their emotional energy into their patients and process the frustration and sadness with their colleagues. Down in the OR, the patient is a presence but not quite a person, and the nurses never meet the families. Most patients arrive unconscious, remain unconscious throughout their time in the OR, and leave the same way. The core relationships are among the team gathered around the patient: the attending surgeon, the residents, the anesthesiologist, the scrub tech, the nurses. Everyone is on a first-name basis, everyone knows everyone else's life story, including the day-to-day minutiae. As for intensity, it could not be greater, but the laserlike focus on the job at hand exists in absolute parallel with a rollicking, often ribald humor that keeps them from cracking under the strain.

John Findley, the psychiatrist on the burn service, told me, "We have to laugh all the time. It's one of the best defense mechanisms there are. Otherwise you're going to sit there and really drink martinis." Findley himself supplies a good share of the yuks. Part court jester, part walking *Physician's Desk Reference*, he can always be depended on for a quip as well as exactly the right solution to a tricky medication issue. What he had to say about the burn team in general applies in spades to the gang in the OR. He started his medical career as a surgeon, so he knows the territory. "We're all kind of toughing it up and we wouldn't admit to that, but that really is the case." The hard shell makes it possible for members of the burn team to survive the long hours and the intensity of the work and the sadness when a patient dies.

Thanks to doctor television shows from *M.A.S.H.* to *ER,* most of us know all too well what happens when the body on the table is out like a light. Surgery is serious business, to be sure, and no one is doing anything other than the best job imaginable, but it can be hard for an outsider to square the tone and content of the interchanges with what ought to be a solemn, or at least sober, setting. Rather than distract them, the music, the joking, the watercooler chat seem to help everyone stay alert and on task. Everyone knows how and when to snap to attention.

About an hour into the preparations, the guest of honor arrives, wheeled into the anteroom through swinging double doors. He is a man in his thirties with a 30 percent burn. Except for his singed face, shiny with Bacitracin, the rest of his burns are hidden beneath soggy brown bandages that cover his right arm and chest. Finally, John Schulz arrives with his residents in tow, fresh from morning rounds upstairs. At last the party can begin.

* * *

The medieval healers who believed that burns were the source of "evil humours" were not so far off the mark. The immune cells that swarm to the scene of a burn release proteolytic enzymes and a host of inflammatory mediators, which in turn attack healthy tissues and fuel the hypermetabolic response, which leads to catabolic destruction. This mob extends wounds through conversion, disturbs blood-clotting mechanisms, and creates a state of generalized chaos that spreads to major organs and affects all body systems. If these are not "evil humours," it is hard to know what is.

Supposed burn toxins can be seen as a more modern incarnation of "evil humours." In light of current knowledge, the notion that these mysterious poisons could and should be "fixed" with tannic acid is hardly less fanciful. The conviction that eschar is a good thing because it protects the surface of an open wound while healing takes place beneath it persisted well into the twentieth century, as did attempts to create eschar or strengthen what had already formed, using everything from paraffin and picric acid to triple dyes. Of course we now know that eschar, whether it forms naturally or is chemically induced, contributes little to the prevention of fluid loss and a great deal to the generation of infection. We also know that far from fixing toxins, it invites the cellular and chemical actors that are as damaging as any "evil humour" imaginable.

The great breakthrough came with the understanding that as long as the wound and the eschar that forms above it are in place, the progressive disease that is a burn will continue to run riot throughout the body. The only way to halt the progression is to surgically remove the burn and close the wound. More than taming shock and treating infection, early excision of the burn itself is what began to save lives, and finally changed not when but whether burn victims died.

In retrospect, it is hard to understand why it took so long to figure that out. Big burns are bad news; when they are both big and deep they are deadly. Yet the effort had always been to protect the burned area until it healed. If the healing took too long, things might be helped along with skin grafts, but the general approach was to wait. Wait until the eschar melted or sloughed off by itself. Wait until the edges of the wound pulled together and closed. Wait until new skin developed under the detritus—dead skin, coagulated blood, steamed or fried or roasted tissue. The result, more often than not, was virulent infection and crippling scars. In some respects it is like so many of the other advances in burn treatment—what seems obvious now was shrouded in mystery. But even when the principle had been established, there were technical problems that had to be solved, not least of which was how to keep patients alive during the surgery, which is often extensive and always bloody.

John Burke was a leading voice in favor of early excision and wound closure, and many others have contributed to its refinement as today's standard of care, but the idea can be traced back to Oliver Cope, who was the lead author of a paper published in 1947 entitled "Expeditious Care of Full-Thickness Burn Wounds by Surgical Excision and Grafting." Instead of waiting for eschar to separate on its own before grafting new skin, an approach associated with "pain and suffering, severe metabolic derangements, multiple septic episodes, and lengthy hospitalization," Cope and his colleagues contended that eschar and all underlying tissue down to the fascia (the papery boundary between the fat layer and muscles) should be surgically removed and the resulting deep wound covered with skin grafts. Although some surgeons tried it with small burns, such radical surgery was considered too risky, especially from the point of view of infection, and was not widely adopted in the immediate postwar years.

Burke began thinking about the problem of infection early in his medical career. "In the middle of my residency, in 1955, I spent a little over a year studying in England, at the Lister Institute, with a man called Professor Sir Ashley Miles, who was interested in the biology of infection, and that work generated an understanding of the methods to prevent surgical wound infections."

Dr. Burke interrupted his autobiographical narrative with a historical digression, back to what he refers to as "the second war [when] penicillin arrived on the scene. The whole surgical world then felt—wishful thinking—that the problem of postoperative surgical infections, which has always been the bane of all surgeons, was going to be solved, because all you'd have to do was give antibiotics and then you'd never have a postoperative surgical infection. But when that was tried clinically, it turned out to have absolutely no effect." By the early 1950s, Burke explained, there was a reaction among surgeons against using antibiotics prophylactically; their failure to reduce the rate of postoperative wound infection, coupled with the expense and the risk of allergic reactions, made them seem like a bad bet. "In addition, surgeons felt that you ought to be able to prevent wound infections by meticulous surgical technique, so they thought that people who used antibiotics were likely to be sloppy surgeons, and that was their problem.

"That didn't make any sense to me because if antibiotics killed bacteria in an infection, I couldn't for the life of me figure out why they wouldn't have an effect on bacteria that fell into the surgical wound when you were operating. So one of the reasons I went to the Lister Institute and Ashley Miles was that he was poking around in this area.

"To make a long story short, what we found out was that all the important biologic activity that generates an infection goes on in the first three hours after the bacteria get there, but all of the people who were giving antibiotics to prevent wound infection gave it after the operation was over. So we said that if you want to prevent an infection, you have to give it before—the antibiotic has to be sitting in the interstitial space when the bacteria get there. . . . And everybody in the world does that now. Nobody doesn't do it."

When Burke returned to Boston and MGH, there was no burn unit nor was he particularly interested in treating burns. "I was a thoracic surgeon, interested in lungs and hearts and things like that. But everybody at Mass General—we're talking now in the '50s and '60s—was really a general surgeon in the sense that those who did operations in the chest also did operations everywhere else, just like general surgeons are supposed to do. So, I still took care of burns when they happened to come up when I was their doc." He fell into burn treatment when the Shriners approached the General about setting up a children's burns hospital and research institute. Oliver Cope was the point man and first chief of staff, but Burke was involved from the early days and succeeded Cope shortly after the Shriners facility opened in Boston.

"Okay, now here's this enormous research institute that the Shriners say they're going to build and what in God's name are we going to do with it that makes some sense and does some good? The basic idea was we had to change the burn system, because we'd been at this since Adam and it didn't work. So it seemed to me that we needed a pretty substantial new look."

Around that same time, Zora Janzekovic, a Yugoslavian surgeon who is the only woman whose name is ever mentioned among the pioneers of burn treatment, published a paper espousing early excision facilitated by a novel surgical approach. Janzekovic argued for removing destroyed tissue because it was not going to heal under any circumstances and its presence merely increased the risk of wound conversion. "We believe that when excision of the devitalized tissue is carried out between the third and fifth days, we rescue the potentially necrotic deeper layer," she wrote. Rather than cut deeply into the burn and beyond, she used a sharp blade to make a series of thin slices, working gradually downward until she reached healthy tissue. This method, which has come to be called tangential excision, was slower and more exacting than the usual approach; because copious bleeding was the signal that the blade had arrived at viable tissue, it was also extremely bloody. But it was much less radical and much less disfiguring than excision down to the fascia, which was the general practice at the time. With her method, it was possible to cover the wound with partial thickness skin grafts, rather than use full thickness grafts, and that in turn made it possible to graft larger burns within a shorter period of time.

Fascial excision certainly gets rid of all the dead tissue. Depending on how deep the burn is, it also removes more than a bit of living tissue. The fat layer is cut away along with all skin layers, and none will grow back. Even when covered with a full-thickness graft, the result is what is called a contour defect, a deep depression or spindly limb, for example. Fat storage moves to existing fat cells elsewhere in the body, resulting in an imbalance in fat distribution that may be grotesque.

Burke tried both tangential and fascial excision, operating from twelve hours to eighteen days postburn, and was able to document that early excision and grafting saved lives and decreased the length of hospital stay. Previously, patients with big burns might be hospitalized for as long as a year while doctors waited for their wounds to "mature" and nurses tried to keep them alive through endless bouts of infection. As long as the burn was there, the "evil humours" kept the patients sick and made them sicker. The conclusion was clear: "The earlier one can excise, the less chemical or physiologic insult the patient will receive from his burned and destroyed skin." Tangential excision was deemed preferable to fascial from a cosmetic point of view.

Burke instituted early excision and grafting at both Mass General and the Shriners Hospital for Children in 1976. The result was a precipitous decline in mortality from an average of 25 percent in 1974, before the practice was routine, to 7 percent between 1979 and 1984. Other hospitals that adopted early excision saw similar declines in mortality. Today, as Dr. Burke is wont to say, "nobody doesn't do it."

* * *

Like a pit crew at the Indy 500, they surround the patient, working in ones and twos on different pieces of the job. Everyone is involved, the nurses, the anesthesiologist and his resident, Schulz and his two, the burn tech. The patient lies inert, the mechanical rise and fall of his chest the only sign of life. Two people roll him onto his side while another props him up with pillows and rolls of toweling held together with surgical tape. Someone else drapes him with blue sheets so only his right arm is exposed. Jennifer Verbesey, the junior resident, starts to take down the dressing, snipping and unwinding yards and yards of Kerlix, leaving a pile on the floor, which one of the circulating nurses collects and deposits in a red biohazard container in the corner. Schulz stands off to the side and confers with Dave Cooke, the senior resident, about the task ahead.

Dan O'Shea's burns were similar to what they are looking at. This guy was burned when he attempted to revive a dying fire in his fireplace by dousing it with charcoal starter and holding a match to it. At least that is the story he told when he was brought in late Monday, conscious and in agony. There is more than a bit of scoffing around the table about his claim, given the ninety-degree heat that has been gripping the Northeast and the fact that no one and nothing else seem to have been burned. Someone pipes up with, "You gotta learn to toss that match from a distance if you're gonna light a fire that way." He had no inhalation injury, but he managed to cook his right arm and a good half of his chest. It looks like he had the presence of mind to "drop and roll" soon enough to confine the damage to one side of his body. Schulz and his crew do not know exactly how deep the burns are, but they are at least partial thickness, probably full thickness in places. What is certain is the arm is worse than the chest, and the hand worst of all. Only when they are actually shaving off the eschar will they know how deep the damage goes.

For all the talk about burn degrees, partial thickness, full thickness, and the like, what ends up mattering is whether a burn will heal on its own or need to be closed with a graft. According to Schulz, "If it needs grafting, we call it third degree."

Skin grafting has a history as long as burn treatment. The ancient Egyptians did it, and so did the Renaissance Europeans, to close wounds, whether caused by fire or steel. Noses, lost on the battlefield or the dueling ground, seemed to be a common site for grafting. Then as now, the best grafts came from somewhere on the patient's own body, rather than from another person or animal. What we now call autologous or autografts usually "take," whereas skin from another person or an animal is invariably rejected, usually sooner than later.

The Italian Gaspare Tagliacozzi is regarded as the father of plastic surgery, clearly a Eurocentric designation. The method of skin grafting he described in *De Curtorum Chirurgia* (1596) came to be known as the "Italian method," though it was cribbed from the much older "Indian method." Before Tagliacozzi, European surgeons would lift, but not detach, a flap of skin from the arm, for example, and sew it to where the nose had been. The patient would be left in that awkward pose until the graft established itself. Rather than bridge the donor and graft sites and disconnect them after the required interval, the Indian and Italian methods used free skin grafts.

Skin grafting has come a long way since then. Surgeons have tools that make it possible to remove strips of skin of uniform depth. They are able to perforate donor skin with a mesh pattern so it can be stretched to cover an area larger than the one from which it was taken or simply provide an escape route for air and fluid that might otherwise collect beneath the graft and keep it from taking. Today's surgeons know how to minimize blood loss and are figuring out how to use the thinnest possible grafts without sacrificing strength and resilience.

Schulz and Dave Cooke map out the grafting strategy: meshed split-thickness graft for the arm, shoulder to elbow, elbow to wrist. The mesh will be forever visible, like the ghost of a fishnet stocking. It will take a lot of physical therapy to restore strength, mobility, and flexibility, but with a bit of fancywork at the elbow and underarm, they ought to bend as they used to once the grafts have healed. They plan to use unmeshed skin for the hand to obtain the best cosmetic result. Schulz expects to be able to get enough for the whole job from this patient's back.

They will not work on the chest at all today. There is only so much you can do with a big burn before time, blood loss, heat loss, and the stress of surgery are too much to ask of a freshly injured body. Instead they will do it in stages, beginning with the worst and the most functionally important area. The arm and hand win out over the chest on both counts. Depending on how the patient is doing postoperatively, they may bring him back in on Friday, or the following Monday at the latest, to peel away his chest burns and see what needs grafting and what might heal on its own. As for why they did not do the excision sooner, it was too late on Monday and, Schulz said, "We don't operate on Tuesdays." Immediate excision, it turns out, means as soon as possible, given the realities of the situation: not just the OR schedule but how stable the patient is, what else is going on that might be a greater threat to recovery, not to mention how soon after the fire the patient arrives. Because the General is a regional referral center, the burn unit frequently gets patients two, three, or more days out from a burn, when the hospital to which they were first taken realizes the injuries need more specialized treatment than it can provide.

Schulz takes over unwrapping as he talks, while Jen moves to the other side of the table to prepare the donor site. She passes a disposable razor over the patient's back to shave off any hair, then tosses it in the small red biohazard bin reserved for used blades, needles, anything sharp. She flushes the area with sterile water, then paints it with a brownish-orange disinfectant.

The tech is filling the turkey basters with epi-sal—a mixture of epinephrine and saline solution. She fits one with a long curved needle and hands it to Jen, then turns to prepare another. Jen slips the needle just beneath the skin of the patient's back and depresses the plunger, withdraws, then reinserts the needle a few inches away and squirts again. When the syringe is empty, she hands it to the tech, who hands her another. Jen works methodically, discharging three, four syringeloads. "Put in as much as it will take," Schulz instructs. He has his eyes on everything that is going on in the room. Gradually the area marked off for harvest takes on a puckered, swollen appearance vividly referred to as *peau d'orange:* skin of an orange. It will be easier to plane off skin from the raised, tense surface, and the epinephrine will retard bleeding.

The two women work together, speaking quietly between themselves. The tech has a businesslike demeanor that is markedly different from the sassy informality of the OR nurses. She exudes a quiet competence as she stands at her station, preparing and delivering requested instruments, replacing blades, handing out towels, sponges, gauze, sutures, whatever is called for from the array of paraphernalia on the table before her. She keeps busy, even when she is not called on to dole out material, by laying out a supply of blades, staple guns, whatever she knows will be needed in quantity. It is hard not to think of a bartender laying out the olives, the twists of lemon and slices of lime, the cherries, the swizzle sticks, setting up a tray of cocktails, sliding them across the bar.

Unlike the nurses who circulate throughout the room, the person standing within the sterile field, gowned and gloved to a fare-thee-well, eyes shielded by safety goggles, nose and mouth covered with a mask, is the keeper of all things sterile, which she receives from the circulators in a neat handoff involving two layers of paper covering, the inside sterile and touched only by her. Sterility is serious business down here, where blood and other vital fluids drip and ooze and squirt from wounds that might already be infected and certainly are open to becoming so. Terri said, "They do things sterilely on the unit, but not to the magnitude we do here. So you have to be conscientious, and if you screw something up, you have to say: 'Don't touch that,' even if it means delaying something or being totally humiliated and feeling like an idiot."

Over on the other side of the sterile field, Schulz and Dave are plumbing the depths of the burn to determine how far down to excise. The burn is not uniformly deep, so they cannot just slice down a measured distance. They do not want to cut off any areas of superficial burn because that will only create a deeper wound that will require grafting instead of healing on its own. On the other hand, they do not want to leave any nonviable tissue, since a graft will not take unless the wound bed is absolutely clean. This is not unusual, but it is exactly the sort of situation tangential excision was designed for. Schulz makes a first cut with a Goulian knife, which looks like a straight razor with a replaceable blade, and eyeballs a setting for the dermatome. They will do the first pass with the powered device because they have a lot of ground to cover, then go back with the hand blade to fine-tune the excision.

Schulz calls Jen over and she does a do-si-do with Dave. He will harvest the graft while she cuts off the dead skin. Schulz hands her the dermatome. It is a big, electric stainless steel device, with a heavy head, a fat handle, and a thick black cord. Unlike the cheese slicer it vaguely resembles, it cuts on a forward push, not a backward drag. The operator has to hold it at the proper angle and stand in the proper position to get a smooth cut that is not too deep and not too shallow. Even in the most experienced hands it is not easy to grip and manipulate the thing all the while holding down the "on" button. It is a mystery why it was not designed for easier use. But when the first drum dermatome was invented in 1936 by Earl C. Padgett and George Hood at the University of Kansas, it was a revolutionary tool, and no one seems to have done much about streamlining the design. A dermatome, strictly speaking, is any device that cuts skin. The various handheld knives—Goulian, Blair, Humby—are dermatomes, but the term is usually reserved for the big powered shavers.

"Have you ever used one of these before?" Schulz asks. "Yes," she replies matter-of-factly and pushes the button. On top of everything else, the thing is infernally noisy. As it moves over the eschar, curls of brown skin yield to a layer of white with pink specks; this is healthy dermis dotted with decapitated capillaries. The tourniquet is keeping the capillaries from spouting blood. After each run, Jen pulls out the eschar with forceps, washes the blade area with a squirt of water, and wipes it with a gauze sponge.

The dermatome removes a lot of tissue in one swipe, but a handheld knife can get into tighter corners, so after a few passes, they transition to the Goulian. Schulz presides, coaxing, coaching, giving directions, sometimes taking up a blade or other instrument himself, either to offer an example or to get the job done faster, though there is no impatience in his voice or his manner. He might urge a surer, swifter hand with: "Come on, you're in the deli, slicing meat. Everyone is waiting for the pastrami." Or he will take a look at an area already excised, feel the surface, and then suggest a deeper cut: "That still looks a bit gnarly." The wound bed has to be absolutely clean, not a shred of dead tissue left, but with the tourniquet inflated, it is not so easy to tell what is dead and what still alive. "You can by sight, with experience, tell what's alive, but the acid test is really the density of capillary supply, so you have to let people bleed a little bit if you want to be sure," Schulz explained. When it looks to Schulz like they have cut off all the dead tissue, he will ask Charlie to deflate the tourniquet for ten seconds so everyone can see what is bleeding and what is not. A bit more excision may be called for.

Schulz is working on the forearm while he keeps a watch on Jen's blade work. He pulls off thin bits of dead skin, using forceps, scissors, his gloved fingers. It comes off like peels of rubber cement. He is in his element. Just returned from his two-week vacation, he takes in the scene, a broad smile rippling his surgical mask. "I wasn't too happy to be back until I got down here."

He takes a look at the hand and does not like what he sees. He asks for a Goulian knife and calls Dave over. He starts shaving, with a firmer grip and a surer hand than either of his residents, but the blade is dull, so he asks for another. The tech clips a new blade into another handle and hands it over without comment, deftly flicks the old blade into the red "sharps" bin, and immediately prepares some more. She just keeps opening packages, loading the knives, lining them up to hand over. Over the course of five minutes, they go through easily a dozen. Who knows, maybe it was just a dull lot; it happens sometimes.

At a certain point, Schulz stops. It has become clear that the subcutaneous tissue is very injured. "There's very little between me and his tendons, and what little there is doesn't look very good. I don't want to keep excising. Let's get this on dressing changes and see what happens, try to let his hand decide what's alive and what's dead, so I don't have to go down and get into his tendons." It is a conservative approach, but the right thing to do in Schulz's judgment. It is, after all, the man's right hand, and he would like to return it to him in working condition. So the best laid plans go the way of the reality beneath the skin. "We usually excise and graft in the same operation," Schulz explains. "But not always. Number one, if we don't have enough donor site, then we would try to use something else, a temporary covering like allograft or Integra, and come back later. Number two, if we're worried about the wound bed." Today they will graft the arm, but stop at the wrist.

Dave is ready for the harvest, which he will do entirely with the dermatome. Cleaned, with a new blade in place, it will provide even slices of epidermis with a thin lining of dermis: skin to be transplanted onto the raw flesh of the man's arm.

This is referred to as a split-thickness skin graft, a relatively new technique in the centuries-long tradition of skin grafting. Until the mid-twentieth century, full-thickness grafts were the standard. Sometimes an expanse of skin was used, but taking a large piece of both epidermal and dermal layers from one part of the body to cover another was, in effect, borrowing a lot from Peter to pay Paul. More frequently small, deep plugs called pinch grafts were planted on the wound bed and left to grow together, a process not dissimilar to hair transplants. The many areas from which the plugs were taken would heal more readily than a single large donor site. Pinch grafts were popular through World War II and were famously used on Clifford Johnson, a young Coast Guardsman and heroic victim of the Cocoanut Grove fire, whose deep burns, estimated between 50 and 60 percent, resulted when he repeatedly returned to the flaming structure to rescue others. Over the course of a year, Newton Browder, the plastic surgeon remembered for the Lund-Browder body surface diagrams, did thousands of pinch grafts using a Gillette blue blade—an ordinary razor—to methodically repave Johnson's body with skin.

It was not until the 1930s that it became possible to remove a slice of skin of even depth and, by extension, a depth shallower than the full thickness of the skin. The invention of the Humby knife in 1936, with its adjustable blade, was quickly followed by Padgett and Hood's drum dermatome. It remained for surgeons to argue about whether split-thickness grafts were preferable to full-thickness ones. Full-thickness grafts look more like normal skin when healed. They are less prone to scarring and contracture. On the other hand, harvesting them produces a deep wound with its own healing problems, and the donor site cannot be reused. Split-thickness grafts may be cosmetically inferior, but the donor site heals readily and can be reharvested if more skin is needed. Nowadays, full-thickness grafts are rarely used to close burn wounds, and the trend is toward increasingly thinner split-thickness grafts.

As Dave pushes the dermatome up the length of the patient's back, it leaves a stark white rectangle, about six inches wide, outlined with tiny red beads. The tech grabs the end of the peel with forceps as it emerges from the top of the dermatome, holding it taut, keeping it from getting tangled up in the works. This is precious material. The aim is a clean-edged strip, as long as possible, of perfectly uniform thickness.

At the end of each pass, the tech frees the strip of harvested skin. It immediately curls in on itself, the beige epidermal surface enclosing the stark white dermis in a rubbery tube. She lays the tissue onto a square of sterile gauze in a stainless steel tray and teases it flat, then squirts it with saline and covers it with another gauze sponge. The tissue must be kept moist—alive—while waiting to be transplanted onto the clean wound bed. When it is placed, it has to be dermis side down. Anyone who has seen it more than once would know which side is up and which down, but Schulz makes a point of emphasizing this all the same. Clearly the error is not unheard-of in the annals of burn surgery.

The tech is an island of calm amidst the hubbub, and though she briefly boogies down with the circulating nurse when an especially danceable tune comes over the boom box, she does not participate in the verbal sparring. Standing beneath the surgical lamps, the hottest spot in a sweltering room, it is a wonder she does not pass out from the heat.

When one of the nurses comes around with a plastic bag filled with ice cubes wrapped in a towel, she lays it on the tech's neck first, holding it there a few moments before moving on to the next wilted colleague. Schulz declines the service, preferring to chug sterile water. As he empties one of the several plastic liter bottles he runs through during a typical session, he takes a look at the label. "Hmm, 'not for human consumption.' But we pour it into people all the time," he says in a mock plaintive tone. "I just don't want to keel over. When I was a resident, I did keel over once." Schulz is more than happy to tell stories on himself. He almost seems to like taking himself down a peg or two. "I was staying up every night with my daughter, who was two years old and had bad ear infections—I had been up all night long for the whole weekend—I walked into a room, and after five minutes of déjà vu I thought, 'I'm going to have a seizure,' and then I hit the deck. By the time I woke up, I was boarded and collared with an IV in and a C-spine collar, because when I went back I hit the wall and then hit the floor, head first. So I spent two days in the hospital."

"You just wanted a day off," someone shoots back.

"Yeah," says Schulz. "The chief of plastic surgery said, 'You could have just mentioned that you weren't feeling well.' Well, humpf." He pauses meaningfully. Everyone knows that as a resident he did not feel he could have, none of them ever does. The key to his rapport with his residents may be that he remembers what it was like and has a healthy skepticism about the pedestal on which surgeons put themselves.

The circulating nurses come and go, moving around the periphery of the room, constantly replenishing supplies, removing waste, wiping up puddles, answering requests, making entries in the nursing computer, pulling up lab values on the patient. They document the entire operation—time into the room, time out of the room, anything that happens during the case, who is there, what went on during the procedure, what the patient was prepped with, what positions he or she was in, and at the end of it all, were the sponge and instrument counts correct—screen upon screen of entries that, along with the anesthesiologist's logs, make up the narrative of the day. Sometimes a circulator ducks into the anteroom, where larger or less frequently used supplies and equipment are kept, or leaves through the swinging doors, to return moments or as long as a half hour later, presumably after lunch or a break.

"If you're at the field directly working with the patient, once the operation gets started, that's considered the sterile field," Terri explained. "We're not sterile; we're there to do all the ancillary stuff. We might not look like we're doing much, but our ears are open all the time. That's probably the secret to OR nursing: You listen to everything, whether it's coming from anesthesia or physicians." In response, they keep up a steady flow—of supplies from the storage cabinets, of music from the boom box, of joshing and lighthearted conversation.

"Hey, John, you wearing your cowboy boots today?" one asks, looking down at the bright white paper gaiters that cover Schulz's legs from foot to knee. Everyone else wears blue paper booties over clogs or sneakers.

"Yup," he says, then offers this slapstick bit. "One time I was wearing them and I was walking down to the cafeteria and whoosh! You know, Mr. Big Surgeon . . . and my feet just totally slid out from under me. There was a collective intake of breath . . . and I just got up as fast as I could, yelled 'Jesus Christ,' and kept on walking." He pauses for two beats. "So I stopped wearing the boots for a while. But now it's been long enough. You know, human beings forget pain . . . kind of." He laughs, the sound bouncing off the tile walls.

Schulz has two different laughs, and he uses both liberally. One is a full-throated chortle that is a real traffic stopper, and the other a slightly more subtle chuckle that punctuates most of what he says, though he uses it most often when he senses he has ventured too close to posturing.

Throughout the operation, there is a fair amount of drifting in and out. At one point, a pharmaceutical company rep comes in wearing green scrubs and an eager smile, which quickly vanishes when he catches sight of the bloody doings on the table. Between what looks for all the world like a rolled roast where an arm ought to be and the gathering gore on the man's back, this is not a sight for the weak of stomach. Ostensibly, the rep has dropped by to see if Schulz needs technical help with the product he is selling, an adhesive designed to hold down grafts and temporary skin substitutes. Except they do not happen to be using it today and Schulz is in no mood for a sales pitch. Right about then, one of the nurses sidles up alongside Schulz, who is holding the excised arm, and in her best Brit accent says, "Pick it up, darling, I want to put down a new tablecloth." The rep excuses himself, clearly appalled by the scene of carnage, and exits stage left. The doors have hardly stopped swinging when the room erupts in laughter. It is a bizarre but apparently common occurrence. "All kinds of people come into this room," Schulz says, shaking his head.

Finally the arm is clean, all dead tissue sliced or peeled or rubbed off. Now it is time to stop the bleeding. Until now, the tourniquet has seen to that, but once it is released, every severed vessel would become a gusher were it not for this next phase. It is a long, tedious, and multistep process, but essential, since bleeding under a graft is the sure way to failure. While Jen holds the arm, the tech hands Dave pad after pad of nonadherent gauze soaked in a solution of thrombin, a blood protein that helps speed clotting. Dave lays the pads on the arm, covering it completely, and then wraps it with a rubber hemostatic bandage. As Dave and Jen sit across from each other on low stools, the arm elevated between them to help the heartward flow of blood, Schulz signals Charlie to remove the tourniquet. In perfect synchrony with this calm moment, Norah Jones's voice wafts through the room. "Don't Know Why" segues into Van Morrison doing "Moondance," and still they sit, waiting out the coagulation cascade. Finally, after about twenty minutes, it is time to undo the bandage and set to work with the Bovie.

Dave wields the yellow penlike wand attached to the console by a dark blue wire, Jen stands ready to irrigate the bloody field so he can see where the blood is coming from. As Dave touches the wand to a gushing blood vessel, it gives off a *zzzzt* and the distinct smell of grilled hamburger fills the room. Jen flushes, Dave zaps, over and over, and then they switch jobs until each bleeding vessel has been burned shut.

At this point, two people are needed to assist so Terri scrubs in, shouldering through the swinging doors with her wet hands raised. The tech hands her a sterile towel and helps her into the cloth gown, a neat little dance that ends with a pirouette that wraps the gown at Terri's waist and allows the tech to tie her in, all without breaching the barrier between the sterile and the not sterile. Finally the tech holds out the latex gloves, turned half inside out so Terri can wiggle her fingers in and then secure the tops over the knit cuff of the gown.

While Dave and Jen are occupied with the arm, the two women work around back dressing the donor site. Harvesting skin for grafting has created a new wound. It is a clean wound and only partial thickness, but it is a wound nonetheless, upping the percentage of the patient's TBSA. That has repercussions all the way down the line: It is a new site for potential infection, a new nexus for the inflammatory response, a metabolically demanding stretch of territory, another factor in figuring out fluid replacement needs. Because it is partial thickness, it is exquisitely painful, just as a superficial second-degree burn is.

At the General, they dress donor sites with scarlet red, as they have for years. This old-fashioned dressing is made from fine-mesh gauze impregnated with lanolin, olive oil, and a red dye that stimulates cell regeneration and speeds healing. Newer materials run the gamut from a plastic film that is, for all intents and purposes, surgical-grade Saran wrap to a nylon mesh coated with pig collagen, which is supposed to hasten healing, and various hydrocolloid wafers, powders, and sheets made from pectin or other gelatins or calcium alginate, which is derived from seaweed. All of these dressings aim to keep the wound moist while establishing a barrier against bacteria. As always, covering the wound also reduces pain. Each dressing has its pluses and minuses, ranging from cost and ease of use to the risk of fluid collecting and infection developing under the adherent dressing. In a way, scarlet red is like silver nitrate. It has been around for a long time; it has long since earned out its research and development costs, if it ever had any, and no one is buying it four-color ads in the surgery journals. But it works fine. The General is a conservative place, slow to innovate. Some say it makes a fetish out of "the MGH way," which tends to be the way they have always done it. What is certain is they do not discard materials and methods that work just for the sake of trying something novel. It has to be more than new; it has to work better than whatever everyone is used to.

While Terri and the tech finish with the donor site, the circulating nurse begins to clean up, putting away equipment no longer needed— the dermatome, the Bovie—and collecting the wet, bloody cloths that litter the scene. The cleanup and setup is an endless dance.

Charlie pokes his masked face over the top of the curtain separating mission control from the sterile field like a neighbor peering over the back fence. He wants the room hotter. At the start, it was unusually cool, but over the course of the long morning he has had the nurses crank up the heat and humidity to steam bath levels, all in the interest of maintaining the patient's temperature. He is, after all, lying there half flayed and totally naked, so it takes a combination of an ambient temperature into the nineties, hot lights over the table, and warmed IV fluids to keep his core temperature close to normal. The nurses on Bigelow 13 get really pissed when a patient comes up from the OR cold, Terri says, so up goes the thermostat and out come the ice bags for everyone else.

Dave stands at the mesher; Schulz and Jen are on the other side, the blanched arm lying between them. Dave runs a strip of skin through the mesher and hands it over to Schulz, who deftly smooths and spreads it over the upper arm, pushing out air bubbles to ensure it makes positive contact with the wound bed. Any air, blood, or other fluid between the graft and the wound bed will prevent graft take and invite infection, though the holes in the meshed skin make that less likely than would be the case with the unbroken expanse of an unmeshed sheet graft. The graft must be meticulously placed so it exactly abuts but does not overlap the viable skin bordering the wound. The wet, rubbery tissue is extremely elastic, and Schulz is able to cover a remarkable amount of the man's biceps. He tries for the fewest possible pieces, the fewest possible seams, since each seam will be a scar. At his signal, Jen staples the edges in place, using a white plastic gizmo that pops out stainless steel staples that would be quite recognizable to anyone who has strung telephone wire across a baseboard. Meshed grafts are usually stapled in place rather than sewn. "The result is essentially the same as using stitches," Schulz explains, "but it's faster," which is especially important when grafting a large area since the less time a burn patient spends in the OR and away from the ICU, the better. (An arm may seem relatively small, but its surface area is actually a bit more than half that of the back.) Some surgeons stick grafts on with tape or, amazingly enough, superglue or the fibrin adhesive the queasy sales rep was touting. Most surgeons sew unmeshed graft, especially on the hands and face.

Dave hands Schulz the next piece, which he lays down just below the first, then snips and fits them with blunt-nosed scissors, smoothing the areas where two grafts join. At a certain point he switches tasks with Jen, letting her fit the puzzle pieces together while he staples. Her hands are more tentative, but that is to be expected, that is why she is here. Schulz is an encouraging coach, standing at her elbow, holding down an edge and occasionally nudging it for a better fit. They work their way down the arm, and finally the sleeve of new skin is ready to be enclosed in a bulky dressing that will stay in place for at least five days.

It takes forever to wrap the arm. Layer after layer of gauze sheeting, then roll upon roll. "You can never use too much Kerlix," Schulz announces as he reaches for another roll. Then he makes a fanfold with yards of gauze sheet and quilts the thick sandwich, taking stitches through the gauze and into the skin to hold it in place. By now, everyone is punch-drunk. The jokes are still coming thick and fast, but it sounds like a comedy convention. Someone gets less than halfway to the punch line and everyone else knows where it is going, so they laugh and keep on working.

The bandaged arm sticks straight out from the man's body. Schulz staples fabric tape along the entire length, pulling it taut to further immobilize the arm. That graft is not going anywhere if he has anything to say about it. This is a ritual he can undoubtedly do in his sleep, but he does not hurry, even though it is well past noon and the room is sweltering and they have all been standing up and concentrating and working like demons for hours, and really all anyone wants to do is get out of there. It can take as long as an hour to apply dressings after surgery, but cutting corners on this last step could undo the entire morning's work. Schulz is not about to risk that.

And then suddenly it is over. The nurses wheel away the instrument tables, take down Charlie's curtain, remove the drapes from the patient, pull out all the blanket rolls and pillows that held him in position. They gently ease him onto his back, and cover him with a foil space blanket. It all has the feel of stagehands striking the set after a performance. Gradually the players drift away, leaving the nurses to do the final cleanup. They have donned yellow plastic aprons to fill the red and yellow bags with what gets tossed and what gets washed. One drags bag after bag out to the anteroom, while the other collects and counts the instruments, then readies them for the autoclave. Once every scrap of evidence of the morning's work has disappeared from the room and everything has been thoroughly scrubbed down, they take a short break before they begin to set up for the next patient, repeating the routine that started their day. The two women work alone, engrossed in their task and their talk, barely noticing when Charlie and his resident wheel the patient out of the room. For a first session in the OR, this was fine. The man has a lot of other issues. He will be back.

* * *

Dan O'Shea got the first of many skin grafts on February 4, four days after he was burned. That is soon enough to qualify as early excision, especially since he spent the better part of his first two days at the General either in the hyperbaric chamber or en route to and from it. His burns were removed and grafted in stages. In the first stage, skin was harvested from his unburned thighs to replace what was destroyed on his left arm, shoulder, and chest. Two days later, his right arm was covered, a week later his hands. Five days after that his left flank donated numerous small pieces of skin to cover spots on his left arm where the first graft had not taken.

For Schulz and the other surgeons, graft failure is not unheard of. Grafts can fail to take for any number of reasons: infection, blood or fluid that has collected under the graft, nutritional deficiency that impairs healing, uneven pressure or shearing on the graft, and sometimes just because. For Jack and Peggy it was another item in the catalogue of setbacks and disasters. They continued to come in every day, but it was exhausting and demoralizing and had never stopped being frightening. There had been his pneumonia and a hepatitis C scare that turned out to be a false alarm. On top of all that was other surgery that had nothing to do with his burns: the tracheostomy, the gastrostomy. Who knew what would go wrong next?

It all was wearing them down. It seemed like every time Dan came up from the OR, he was wrapped even more tightly in gauze. One time his left arm stuck straight out from his body, held in position with an ugly brown fiberglass splint. Two days after that it was his right arm, so now both arms were stretched out like he was on the Cross. Less than two weeks later, he came up with both hands bandaged like two cantaloupe melons. It was hard to imagine how the nurses were able to change his other dressings and do all the things they needed to without disturbing the new grafts. But somehow they managed to take care of Dan and have patience to spare for Peggy and Jack. By the time the tops of his ears were grafted on March 2, Dan was into his second month in intensive care, and Peggy was sure the nurses were growing tired of seeing them, the doctors tired of answering their questions.

But there was some good news: Now that his wounds were all covered he did not need to be in the BCNU. When he was brought up from the OR that early March afternoon, he was installed in room 24. Finally they could sit by his side without a plastic wall between them. Somehow that seemed like progress. He was still on the respirator, still in critical condition, but they could see he was beginning to heal, and they could only pray that the wave on wave of crises had come to an end.

CHAPTER ELEVEN

Healing

Tom Parent's left arm was not healing. Every morning, Carl would take down his dressings, scrape off the layer of protein-rich ooze that had accumulated overnight, and then talk him through the agonizing wait for the "A Team." And every morning, the doctors would look at the raw, red expanse of his forearm and shake their heads. It was early June, more than a week since he had been burned, and they expected to see some sign of spontaneous healing. It was, after all, a partial-thickness burn—mostly superficial, though it extended from just below Tom's elbow all the way to his fingertips. It was the sort of burn they could reasonably expect to heal without grafting. Daily debridement and bandages soaked in silver nitrate ought to have kept the wound bed clean and free of infection. Tom was in generally good health and appeared to be well nourished. They had managed to stop the previous week's infection dead in its tracks. They decided to give it another few days, but if they hit the two-week mark and still saw no sign of skin regeneration, they would have to graft the arm and hand. The hand concerned them especially. The longer it took to heal, the greater the chance Tom would end up with hypertrophic scarring, the tough, raised, ropelike scars that can contort the joints, forcing fingers to seize up or binding them together with unyielding webs, rendering the hand useless. Tom needed both hands. He needed them professionally and psychologically. So if it took grafting to give him back his hand, then grafting it would be.

The news threw both Tom and Nancy, though he exhibited bravado while she swallowed her disappointment. "They had already told me what to expect, and it wasn't something I was looking forward to. But, you know, that's just one more trial. What am I going to say? No?" Tom flexed his fingers, stretching his palm wide, then made a tight fist, as though to remind himself what had been at stake.

Approaching two years after he was burned, he still has a small collection of tics, reflexes almost, that date from his ordeal. Some have fallen away, like the panic he felt every time he opened his eyes to the morning light behind the orange curtains in his hospital room. It looked like fire, and no matter how much he knew he was in a safe place, it still made him anxious the whole time he was there. Others have persisted. Whenever Nancy opens a package wrapped in cellophane, the sound reminds him of the crackling of fire. "He'd be in the family room and I'd be in the kitchen and I'd hear a very tense voice say, 'What is that? What is that?' It's not so bad anymore, but I would say for the whole first year, if I was getting ready to do something, I'd have to say, 'I'm going to open some cellophane now. Okay, Tom?' or 'I'm putting something on the stove now. It's just water, and I'm just gonna run to the bathroom for a minute, okay?' Because that would make him berserk."

"It still makes me crazy," Tom admitted. "I don't like an unattended stove." In fact, it was a year before Tom even approached the stove and screwed up the courage to fry himself an egg. "When Tom walks through the kitchen near the stove, his left arm is behind his back. And he doesn't even notice it," Nancy told me. Emotional healing takes a long time, longer than the physical healing, and there is no surgery to speed its progress.

Nancy was on her way in to see Tom one morning when the unit nutritionist stopped her. Tom was on solids, had been since day one, and he was among the rare patients on Bigelow 13 who could order from a menu and get something resembling real food. He tended to order carefully, though, because he figured he was not burning many calories lying in bed all day. He did not want to come home from the hospital looking like a blimp. So it surprised Nancy when the nutritionist told her she was putting Tom on a liquid nutritional supplement. Here was Tom holding back and now they wanted to up his daily calories. Nancy did not get it. "Then she told me that with a burn, you have to eat a tremendous number of calories, your body has to have calories to heal itself, and he was not eating enough.' So I said, 'You know what? Just tell him that.' He was in heaven." Tom started eating everything that was not nailed down.

Tom laughed at the memory. "Of all the bad things that had happened, this was such a good thing, because I could have everything and anything." He would send Nancy down to the snack bar in the lobby for ice cream smoothies. "When I could get around, I'd go down with her, and that was my big trip, my big excitement for the day."

It must have worked because the day before he was scheduled for grafting, the sign everyone had been looking for finally appeared. "The team came in and looked, and they saw little skin buds where I was going to have the grafts." Tiny pearl-white islands of epithelium—the precursor of new skin—dotted the sea of red along the top of Tom's forearm and the back of his hand. They canceled the surgery and left it to Tom's body to repair the wound.

* * *

The healing process begins almost immediately after a burn, while the body is in the throes of its riotous reaction. Indeed, healing is intricately bound up with the inflammatory response, and it should be no surprise to find the familiar double-edged sword: "Paradoxically, the same systemic and potentially damaging inflammatory response triggered by extensive thermal destruction of the cutaneous envelope will provide the basic stimulus for reconstitution of the skin barrier." Although there would be no healing without inflammation, the general chaos immediately following a burn creates an environment that is not conducive to healing.

Healing is all about timing and the delicate balance between destruction and construction. The most intense activity takes place in the first three weeks after a burn, but significant changes at the site of a wound continue for at least a year. It begins with a tightly controlled sequence of cascades, those complex chain reactions that summon and switch on or off a procession of specialized cells and chemical mediators.

Wound healing can be divided into four stages. They are not strictly chronological. Rather, they often overlap and at times occur simultaneously. The first stage is coagulation, the body's attempt to stanch bleeding and fluid loss at the site of injury by constricting blood vessels and sealing the wound. Players in the coagulation cascade include platelets, thrombin, and fibrin. In deep burns, heat contributes to sealing off the wound through eschar formation, coagulation of blood, and destruction of blood vessels.

The second stage of healing is dedicated to cleaning and preparation of a wound bed on which new skin can grow. It coincides with and is driven by the inflammatory response. It begins with the arrival of neutrophils, which assemble a mop-up squad of proteases and macrophages. This crew begins wound debridement long before a healer's hand touches the wound. It attacks bacteria and works to dissolve dead tissue and cellular debris, including the eschar that sealed the wound. The neutrophils also signal the release of the first of a host of growth factors, specialized proteins that control skin and blood vessel regeneration. These proteins give cells their marching orders. Their release may begin early in the inflammatory phase, but growth factors continue to be present and active until healing is completed, a year or more later. The interplay of growth factors and proteases is also highly complex. Some are mortal enemies; others work in an uneasy alliance. Some switch sides midway through healing. For example, early on, metalloprotease, which is present in the fluid that oozes from the wound, interferes with healing by blocking growth factor activity. That is why oozing wounds must be thoroughly debrided every day. Late in the healing process, collagenase balances collagen synthesis to keep it from getting out of hand and producing a scar that is too big, too tough, and too thick. Much of what goes on is mysterious, or at the very least what has been called ambivalent. "The same cellular and molecular structures aimed at destroying and eliminating wounded tissue will now change functions and will proceed to a very opposite task . . . the build up of a new connective tissue . . . first as a provisory assurance of tissue continuity and later as a permanent fibrous bridge between the edges of the wound (scarring)." The complexity of the interaction is part of the reason why infusing or applying or in some other way introducing one or even a handful of growth factors into a wound to speed or improve healing has not revolutionized burn treatment.

Close on the heels of the cleaning crew come fibroblasts, immature cells that are nonetheless highly productive and enormously versatile. These cells migrate into the wound from healthy tissue in the vicinity and direct the restoration of the broken skin barrier. They provide many of the raw materials and issue the orders for the third stage: reconstruction.

During reconstruction, the wound is closed with new skin or a scar. Strictly speaking, this is only one stage in a lengthy process, but wound closure is what most of us think of as healing. Although the greatest visible change takes place during this stage, it takes only about a week or two, compared to the minimum of twelve months that it takes for a scar to mature and the inflammatory response to finally end.

What is reconstructed has a lot to do with what was destroyed. The shallower and smaller the wound, the simpler the process and the more quickly it is accomplished. The deeper and larger the wound, the more likely it is to require surgical intervention.

The human body closes wounds in three different ways: by reepithelialization, scar formation, and contraction. Some wounds may close by only one or two, but most heal by a combination of all three modes. Contraction is the least complex and contributes the least to closing any but the smallest wounds, but because it is often confused with contracture—the crippling and disfiguring traction caused when scar tissue shrinks and pulls on surrounding skin and underlying structures—it warrants explanation first. Contraction is a purely mechanical occurrence. That is, no new tissue is manufactured; the edges of a wound simply pull toward the center, reducing the overall area of the wound. This does not do much for a large wound. The smaller the wound, however, the closer the edges are to begin with, so it is possible for a very small wound, even a very deep one, to close by contraction. The skin removed for pinch grafts is an example. Contraction alone can close wounds the size of a fifty-cent piece; anything larger needs help from the other natural types of healing or surgical intervention.

Reepithelialization and scar formation are separate modes of healing, but their mechanisms are intertwined in quite an amazing way. Reepithelialization rebuilds the epidermis; scar formation takes place in the dermis and is the endpoint of a long series of cellular and biochemical events that begins by signaling and directing reepithelialization and, where necessary, providing the foundation on which new epidermis can be built. Of the two skin layers, only the epidermis has the capacity to replicate itself exactly. It does so through the most primitive type of reproduction: multiplication by simple cell division. Dermis can be *replaced*, but like many replacements, what you get is not as good as what you had.

Compared to dermis, epidermis is rather simple stuff. Its job is to keep moisture in and invaders out. Pigment cells scattered through it provide color to the skin and sentries called Langerhans cells act as an early warning system in case of invasion, but neither contributes to the structure of the epidermis. Structurally, it is a five-layer cake in perpetual motion consisting of a single type of cell that changes shape and form as it matures. Keratinocytes begin life as epithelia, immature cells that eternally proliferate by division at the basal level, the deepest of the five epidermal strata. They mature as they migrate upward, populating each of the other strata in turn, until they arrive at the surface, the stratum corneum. By then they are dead, flat, keratin-filled scales, which flake off to make room for the cells arriving beneath them. It takes about three weeks for a cell to travel from the basal level to the outermost surface. This cycle of self-renewal is continuous, even in the absence of a burn or other injury to the epidermis.

In contrast, the dermis is far more complex in both structure and function, so complex, in fact, that it cannot replicate itself if damaged or destroyed. It can only be repaired or replaced by scar tissue, which lacks the architecture of the dermal layer and performs some, but not all, of its functions. The dermis provides a strong and resilient layer that is both a foundation for the epidermis and protection from trauma for what lies beneath it. It houses blood and lymph vessels, and nerve fibers. Most significantly for healing, the dermis is a factory for tissue repair, supplying both the instructions and the replacement parts.

The cell responsible for most of this is the fibroblast. Fibroblasts manufacture the connective tissue that comprises the dermal landscape: fibrous collagen and elastin laid out on a matrix of glycosaminoglycans (GAG), complex molecules made from polysaccharides bound to amino acids, a sort of marriage of sugar and protein. Fibroblasts also secrete growth factors and other signaling proteins involved in the maintenance, repair, and replacement of skin components.

Wandering through the landscape are other cells that play a part in various stages of the life and death of the skin: platelets (coagulation), leukocytes and macrophages (inflammatory response), and endothelial cells (blood vessel lining). Structures lined with epithelial cells poke down from the epidermis into the dermis like tiny cul-de-sacs, their openings on the surface. These epidermal appendages—the hair follicles and the sweat and oil glands—are crucial to the healing of wounds that extend into the dermis.

Between the epidermis and the dermis lies the basement membrane, a thin, fibrous sheet that both separates and attaches the two skin layers with a gluelike substance called fibronectin. The dermal-epidermal junction is a wavy border, rather than the meeting of two flat planes, and a lot of communication and activity take place across the divide.

The repair of sunburn and other superficial burns that go no deeper than the basal layer is a slightly accelerated version of the normal epithelial self-renewal. Inflammation following the injury signals the epithelial cells to divide more rapidly and begin their upward journey, reconstructing the full five strata of the epidermal layer as they go. The neoepidermis is fragile at first; it takes a week or so to thicken and toughen, and for its color to match its surroundings. Still, burns that involve only the epidermis heal quickly, and they do so exclusively by reepithelialization. There is no open wound to contract and there are no scars, which are a product of the dermal layer.

When a burn breaches the basement membrane, it is considered an open wound. A partial thickness burn can still close by reepithelialization, but the basal stratum has been destroyed, so it can no longer make new epithelium. Instead, epithelial cells migrate in from healthy tissue beyond the edge of the wound and up from the lining of the epidermal appendages, remnants of which survive in the dermis. This astonishing march is orchestrated by the fibroblasts, which begin by preparing the ground the epithelial cells will march across.

Fibroblasts arriving on the scene after an injury start making the GAG matrix and laying down collagen bundles on it. They also secrete growth factors that stimulate the construction of a new network of blood vessels, using resident endothelial cells. The new blood vessels loop their way through the matrix, bringing a rich blood supply to the area to feed the reconstruction. The tops of the loops stick up through the gel-like matrix, resulting in a pebbly surface. This is granulation tissue, a cell- and blood-rich environment that serves as a temporary surface for restoration of the epidermal layer before toughening into scar tissue.

Healing requires a clean, viable wound bed. That is what the cellular mop-up squad is aiming for, what excision of dead tissue is all about, why eschar must be removed and exudate debrided. It is also the purpose of granulation tissue. Granulation tissue is deep red and moist, and it bleeds easily. Nancy Parent was surprised to learn that blood was a sign that Tom's burns had begun to heal. "The first day we started to see them bleed, everyone was so excited." Tom laughed. "Yeah, that was a big day, the day I started to bleed."

At the same time as they are managing the construction of granulation tissue, the fibroblasts send out the signal for the mass migration to begin. Epithelial cells march in from the edge of the wound and climb up out of the epidermal appendages to make their way across the well-lubricated surface of the granulation tissue, at the rate of about 1 millimeter a day. Dividing and multiplying as they go, they form into tiny islands—the pearly "buds" that appeared on Tom Parent's arm just in time to save him from surgery. The islands expand and merge with other islands until the entire wound is covered with a thin layer of epidermis. Once the wound is covered, the epithelia get the signal to switch from lateral movement and begin rebuilding the epidermal strata by journeying upward.

All the while fibroblasts have continued to lay down collagen to strengthen the bridge across the divide. In a sense, it is a race between reepithelialization and scar formation. If the epidermis can be rebuilt and the wound closed before granulation tissue turns to scar tissue, the wound will heal without a scar. If not, there may be a combination of new epidermis and scars. And if reepithelialization takes too long, scars will predominate. Contraction is also taking place, but it does not choose sides in the competition, since a somewhat smaller wound benefits both teams.

Wound depth can delay or prevent reepithelialization. The deeper into the dermis a burn extends, the fewer appendages there are and the scarcer the supply of epithelial cells. Full-thickness burns destroy the entire dermal layer and obliterate all traces of the appendages. They cannot, therefore, close spontaneously at all, unless they are so small that contraction alone will do it. Rather, all burned tissue must be removed and the wound closed surgically. If the skin is very lax, the edges might be pulled and stitched together. If there is not enough skin to bridge the gap, the wound may be closed with a piece of healthy skin from elsewhere on the body (an autograft) or an artificial skin substitute, or a combination of the two.

Aside from size, a number of other factors influence how quickly a burn will heal. In general, burns take longer to heal than wounds of similar size from other causes. The degree and type of cell destruction, the release of inflammatory mediators, the presence of free radicals and proteases, and the shortage of blood and oxygen in and around the wound all work against tissue repair. Protracted or extreme hypermetabolism and its aftereffects are a huge obstacle to healing. Healing will be slow, if it occurs at all, in a patient who is malnourished or suffering from any number of diseases—most notably diabetes. Drugs or anything else that suppresses the immune system even more than the burn itself will also interfere with healing. Some topical agents slow healing and some wound care damages delicate new tissue. Local or systemic infection is another enemy of healing, as is a wound bed that is dry or littered with dead skin and other debris. Burns that cross joints may repeatedly reopen if not splinted or immobilized in some other way. Even stress hormones, which are released when pain is inadequately controlled, can put the brakes on healing.

The longer it takes for a wound to close, the more likely it will do so with an abnormal scar, one that is either exuberantly overgrown (hypertrophic) or so thin and fragile that it repeatedly opens, resulting in a chronic wound and, potentially, an invasive form of skin cancer. It is a rule of thumb that burns that will take longer than twenty-one days to close spontaneously should be excised and covered surgically. The dicey part is knowing which burns those are. There is no debate about the ones that are, at one extreme, superficial partial thickness and likely to heal before the cutoff date; at the other extreme, it is easy to determine that full-thickness burns need surgical closure, and the sooner the better. It is the ones in between that are problematic. No one wants to subject a burn patient to unnecessary surgery, which will deepen the wound and require the use of healthy skin for grafting, producing yet another wound in the process. On the other hand, waiting too long for a burn to heal spontaneously can have far-reaching negative consequences, beginning with the persistence of the hypermetabolic state, the risk of wound conversion, and the perils of infection, and extending into a future of crippling and disfiguring scars.

Once the wound bed has been paved with collagen bundles, the fourth and final stage of healing begins. Scar maturation, or remodeling, takes a year or more, as growth factors and collagenase dance a minuet between collagen synthesis and collagen breakdown. Over time, the raised, angry scar becomes flatter and paler as the supplemental blood supply brought into the granulation tissue recedes and inflammation subsides. If all goes well, the result will be a "normal" scar—a tough protective surface that is less elastic than skin, devoid of hair and the oil glands that keep it supple, not exactly the same color as adjoining skin. It is not perfect, but it does the job.

Given how common scars are, it is surprising how little medical science understands about them. One thing we do know is that scarring is unique to mammals, though the reasons for that are lost in the evolutionary past. Another thing we know, thanks to innovative surgical techniques to correct life-threatening defects in utero, is that the human fetus heals without scars, but that ability is lost at birth. Scars are difficult to study because no other animal suffers the kind of abnormal scarring that humans do.

Hypertrophy is the major scar abnormality. It is the bane of caregivers, from surgeons to rehabilitation therapists, and survivors alike. Hypertrophic scars do not flatten, soften, and pale. Instead, they form thick, raised, tough knots of tissue that continue to feel and act inflamed: hot, red, and itchy or even painful. A keloid is a particularly extreme type of hypertrophic scar that extends beyond the boundaries of the wound, encroaching on surrounding tissue in a way that is almost like a tumor. Keloids can be huge, grotesque, and depending on their location, a severe impediment. For reasons not well understood, the tendency to form keloids runs in families and is more common in some racial groups—blacks and Asians—than others. Although some tribal peoples in Africa have turned this racial anomaly into an art form with decorative scarring, hypertrophic and keloid scar formation in burn victims is not a welcome phenomenon.

Aside from the observation that the likelihood of hypertrophic scarring increases with the length of time it takes a wound to close, no one really understands the underlying cause of these problem scars. Some researchers have found an overabundance of transforming growth factor-beta 1 in hypertrophic scars, suggesting a possible causative role, but this line of investigation is in its infancy and its meaning unclear. It is possible to see microscopic structural differences between a normal and a hypertrophic scar. In a normal scar, the collagen is laid down in a tidy parallel arrangement. In a hypertrophic scar, all is chaos: clumps piled on top of each other helter-skelter, whorls instead of straight lines, and generally too much collagen for the space it is intended to fill. Something has gone wrong with the maturation phase: the balance between collagen synthesis and degradation has gone awry, the orientation instructions have been misunderstood, signals have been blocked or overamplified.

Although scar shrinkage is normal, it is problematic. Scars shrink considerably during the maturation period. At the very least, a mature scar will be tighter than surrounding skin. A fragile scar will continually open as it shrinks and pulls away. A very tough one will fight a tug-of-war, producing a contracture that may range from the uncomfortable to the crippling. Hypertrophic scars are particularly prone to contracture. The best defense against scar shrinkage and contracture is physical therapy, which employs a combination of splinting, traction, and both passive and active stretching. It starts at the same time as healing—as soon after the burn as possible—and continues long after the patient has left the hospital. When physical therapy cannot prevent contraction, surgical scar revision may be necessary. Contracture is a constant threat to full recovery, both cosmetically and functionally.

Preventing scar contracture or correcting it surgically is necessary in adults, but it is absolutely crucial in children. When children's bones lengthen, their muscles, tendons, and healthy skin grow to accommodate the bones. Scars cannot match that growth. Instead they tighten and pull against the skin they are attached to, exerting powerful traction on underlying muscle, connective tissue, and skeletal structures. Years after surviving their injuries and enduring their recovery, badly burned children suffer anew from this cruel impediment.

* * *

Tom Parent loved his Jobst pressure sleeve; Dan O'Shea did not want to have anything to do with the thing. That pretty much sums up the divide within the burn care community about the scar-reducing power of pressure garments and face masks. Some people swear by the skin-tight elastic garments and the custom-made clear plastic face masks, believing that steady and constant pressure on scars as they mature results in a flatter, smoother, paler, less visible scar. Others contend there is no good science to support that belief, and the discomfort and expense—both of them considerable—are simply not justified by the result.

Although using casts and pressure dressings to control swelling was an old idea, pressure garments for scar control came into vogue only in the 1970s, when it was noted that the scars on a burn and trauma patient whose leg had been broken as well as burned seemed to be flatter and softer where covered by the cast than elsewhere on his body. Soon the idea of compression therapy took hold, though the scientific evidence to support its long-term benefits is slim. Its advocates claim it reorganizes the chaotic jumble of collagen seen in hypertrophic scars, enforcing a parallel arrangement that shows up on the surface as a flatter, smoother scar. Others argue that in the absence of a clinical trial comparing compression therapy with unmediated scar maturation, such claims are purely speculative or based on the flimsy science of anecdote. Some in the burn community concede it is wishful thinking as much as anything else, the desire to do something—anything—for their patients, even if it only makes them feel better about a problem that vexed Hildanus and Paré and that, centuries later, no one has yet solved.

Jobst is the big name in the custom-fit business, but there are other companies that produce made-to-measure garments—from gloves to full-body suits. The fabric is a heavy elastic, closer to your grandmother's girdle than to Lance Armstrong's bike shorts. Clear plastic masks are constructed to fit the individual patient's face, made to order from old-fashioned moulage molds or high-tech laser scans and computer simulations. They have to be revised or replaced periodically as the scars shrink and facial contours change. None of this comes cheap. There are other, less expensive compression products, among them elastic tubular support bandages and wrappings of the sort available at any drugstore and used for everything from tennis elbow to carpal tunnel syndrome; silicone and silicone-elastomer sheeting, for areas not easily covered by a compression garment; and stretchy rubberized tape that sticks to itself and can be wrapped more or less tightly and easily removed.

It was not quite a year and a half after Tom Parent was burned—Nancy could tell you. It was exactly seventeen months. She could tell you the hours too. She had gotten over the guilt, mostly, but she still divided time between *before* and *after*. They were sitting in the living room of their Cape Cod house. Tom had just taken down the screens and stowed them away for the winter, the heavy elastic armband that hooked around his thumb and ran up to his elbow showing its age. "This is my work Jobst, the one I wear on weekends, for chores around the house." He also had a dress—one that was not quite so tattered. "I've still got six more months to go and then I'll start to wean myself off. Dr. Ryan said it becomes like a security blanket. And it is. Without it, I feel very vulnerable, very unprotected."

Tom looked like a man who was completely healed, but looks can be deceiving. "I don't have full strength in this hand," he said, flexing the fingers on his right hand. "If I make a fist, my knuckles still hurt. When my hand gets cold, it's painful." He had a hunk of putty he squeezed to keep it limber. "When I drive, I'll stretch my thumb back. If I don't stretch it I'll lose that movement. And it's extremely tight down here, on the pad where my thumb is. They call it banding. So I still work it. And when I get tired or when it gets cold or something, it'll cramp up." Scar tissue is never as good as skin, never will be. It has no blood supply, no pigment, no oil or sweat glands. It is fragile, prone to injury, more vulnerable to sunburn than normal skin, and incapable of healing in a normal way. When Tom is out in the sun he has to keep his scars covered or slathered with high-octane sunscreen, something he will have to do for the rest of his life. It comes with the territory, he figures. He knows he is lucky his burns were not worse.

As for Dan O'Shea, he was not about to wear the heavy, sweaty body armor for the prescribed twenty-three hours a day. "For a while I was wearing them and I stopped, because even at Mass General they weren't recommending them 100 percent. They were saying: 'Some people say yes it works, some people say no.' So I went and got them, but they wanted me to wear them 24/7 for months. It was just not comfortable for me, so I ultimately chose not to continue." Dan was coping with a lot more than his scars, and he did not have a partner like Nancy Parent to cheer him on.

"I just said: 'Forget it.' Because it was just another thing, you know. Oh god, there were stockings all the way up my arm, and on my hands, and I was like at that point: 'I gotta do this?' I mean, something like that can really get you . . . you know, upset."

Mary-Liz Bilodeau says that more than half of the burn patients leave Bigelow 13 with pressure garments. John Schulz is not inclined to push it. "Long-term? I don't think it makes a difference. In the short term I think it does," if for no other reason than the protection, both psychological and physical, it provides. That is one of the generally agreed on payoffs of pressure bandages, even among those who doubt they work for scar control. They give patients a layer of protection once their burns are no longer covered by dressings. They also seem to help control the itch.

Burn survivors who have undergone agonizing procedures and arduous rehabilitation talk about the itch as the worst thing of all. It keeps them awake at night and distracts them every hour of the day. Scars itch fiercely during the year or so it takes them to mature, and the itching may never go away.

Scratching invites infection and breakdown of the still-fragile tissue, which may blister and ooze. Hypertrophic scars itch even more than normal ones. Pressure bandages seem to tame itching somewhat. Moisturizing creams help a bit, as do topical anti-itch ointments and antihistamines. Many people swear by aloe vera. Others recommend cool compresses and soft clothing. The itch is a constant topic at burn clinics and in survivor support groups. No one really knows how to make it stop.

There is no question that a scar—even a normal scar—will never be as good as the skin it replaces. It will never act, look, or feel quite right. Enormous strides have been made in burn treatment. People who would never have survived twenty-five years ago walk out of the burn unit and into the rest of their lives. But no one has yet figured out how to keep scars from forming. Even when they do not interfere with physical functioning, scars are always there, for the survivor if not the rest of the world to see. They are what sets burn survivors off from other people, what makes it impossible to ever forget what happened. For that reason above all others, the scar problem remains the most important unsolved challenge in burn care and research.

* * *

For Peggy and Jack O'Shea, every day reminded them of the first day, the news bulletin on the radio, the phone call, the frantic drive into the city. If Nancy Parent could see progress, the O'Sheas could see only set-backs. If Nancy could talk to Tom and have him answer, could see him interacting with the nurses, the techs, the housekeepers, every human being who came within earshot, Peggy and Jack could only pray and keep up a one-way conversation with their son. As for Dan? Well, it was just a huge blank. He went to sleep one night and woke to find himself wrapped in soggy bandages and unable to speak. Ask him today what his thoughts were and he will not be able to tell you. Because the truth is, for burn patients, there is no moment of truth. Coming to know what happened is a gradual process, one that may have to be repeated over and over until they fully understand the enormity of what Dan calls his "new reality."

That is true of the accidental victims and the would-be suicides, the young and the old, those who spend a week in the twilight zone and those like Dan who remain unconscious for a month or more. A lot of it has to do with the drugs they are given, which produce both sedation and amnesia. It takes awhile to clear them from the system.

"It's not the same as waking up from a nap," Nancy Giese told me. "You don't just wake up and you know where you are and you're coherent and you recognize people and faces and voices. There are just too many medications on board." A better analogy, she says, is someone who has had one too many drinks. "You can have a coherent conversation with them and then the next day they'll have no recollection of that conversation. So they might need to hear something ten, twenty times before it makes sense to them. That's okay; we just keep repeating." Nancy is matter-of-fact about her own patience. She is one of the most centered people you will ever meet, and it is easy to imagine her saying the same thing for days on end without a hint of exasperation.

There is no designated informer on Bigelow 13, and no formula for how and when and what patients are told about how they ended up there. Sensitivity to each individual's emotional and cognitive state plays a big part. "If people aren't asking what happened to me, or if a loved one has died and they're not asking, we don't bring it up. When people want to know, they'll ask. Some people aren't ready to hear. You know, reality is very hard. So we kind of wait until people give us a sign that they want to know," Nancy said.

Bob Droste has watched thirty years' worth of burn patients regain consciousness in the BCNUs. "They fade back into the picture. It's not out one day and awake the next. Even when we cut back on the medication, it doesn't happen that way. They sort of fade in, they know a little bit more, a little bit more, a little bit more. A few times they have some distorted memories of what happened in here, but most people, when they move down the hall, won't remember being in here at all."

"Emergence from sedation is usually gradual enough that the patient kind of assimilates the information gradually," John Schulz explained. "Then usually they put it together. First there's just 'I'm awake again,' and then there's 'Oh my god, what's happened to me?'" He remembers a patient who cried for days. Bob Droste remembers another one whose awakening was a recurring nightmare. The man had survived a fiery multicar pileup, but just barely. He had massive and deep burns, especially on his lower body, and both feet had been amputated. "He was very sick, and he woke up very slowly," Bob told me. "And one day he looks down and he says, 'Oh fuck! I don't have any feet!' Well, he said the same thing many times over the next few days because he rediscovered the same thing over and over."

It was not much different for Dan. His hospital record says he regained consciousness while still in the BCNU. Mike Wilson spoke to him, so did his parents, and so did his close friend, a priest he had known since he was in college. But that is not the way Dan remembers it.

"When I came to consciousness, I had my own room. I was pretty much hooked up to every machine in the hospital, and my arms were out like this." He extended his arms like a crucifix. "And I was breathing on a respirator and I couldn't speak, and that's how I came to consciousness—not knowing anything about anything that had happened. My father and my mother were there, and my father kind of gently described to me what had happened and the situation I was in. And that became the strange new reality of my life."

Everyone was concerned that Dan would have an "Oh my god" moment. His movie-star good looks mean a lot to him. His father hesitates to call it vanity, but Peggy is blunt about it. He is vain; there are no two ways about it. Designer label clothes, a gym-fit body, the handsome face with which he had always met the world—how would he cope with the patchwork repair of his skin, the mottled colors, the raised seams criss-crossing his delts and pecs? There were times, Mike Wilson remembers, "when Dan's parents really struggled with wanting him to survive but thinking he would not want to survive being at all flawed or scarred."

It is a myth that there are no mirrors in a burn unit. In fact, there is a mirror in every room on Bigelow 13, and that is true of any burn center you might walk into. But Peggy and Jack begged Mike to cover the one in Dan's room until he had a chance to get used to what had happened. "I don't do it often," Mike told me, "and then only when the family or loved ones convince me that there is a need to. But I covered his mirror in room 24. We really did not want him to see himself in an unconscious way," without preparation, without someone there with him. Looking in that mirror is something patients do with a lot of support: from the nurses, the techs, pastoral counselors, their families. John Findley works intensively with the self-igniters and any patient who has substance abuse issues or is depressed.

When he talks to them, he pulls no punches. "A lot of times people say to me, 'Wow, I can't believe how blunt you are.' But one of the things I like about psychiatry is I get to ask things that nobody else asks, all kinds of things that are taboo culturally to ask. The reason I left surgery was that it wasn't invasive enough," he said with a sly smile. "My job is not so much to have patients come to terms, but to be sure that they are safe. So I need to know whether they still have the intent to kill themselves. There is nothing to be gained by walking on eggshells. I just go in and say: 'What's the story? What did you do, why did you do it, are you thinking of doing it again?' That has to be done right away—the moment they're able to, the moment they're not delirious or suffering from acute confusional syndrome—because you can't let somebody try to kill themselves and not know if they're waiting for you to walk out the door so they can try it again."

Regardless of the circumstances in which a person was burned, coming to terms with the enormity of it all, with the disfigurement and disability, is a long-term proposition, much of it taking place after a patient leaves Bigelow 13. It begins there, continues at Spaulding or another rehabilitation hospital, which is the next stop for many burn patients, and in the months and years after a patient returns home, to work, to social interactions. Bob Droste likened it to the familiar stages of grief. "It really takes a long time to go through all of the stages. As they start to understand what it is that's happened to them, the body image problems usually come long after the initial wake-up. They don't usually start much of that while they're here. There's no endpoint in sight so they can't start really processing most of that." The progress from the unit to room 24 and then down the hall is just the beginning of a long road. "The scarring issues usually arise after they have left here. When they're here we have to tell them that this is not the final product. So you can't resolve and come to grips with it because it isn't what you're going to have to come to grips with. The inability to do things, again, you can't come to grips with that while you're here because you don't know what you'll be able to do when you're done with rehab. It can be depressing, though," Bob acknowledged.

"The first time out of bed, even if we lift them out and put them in a chair, it's exercise for the nurse, not for the patient, but it's good for their heads. And when they're well enough, it's really good to roll them down to the windows so they can look out and see something besides their room. That is also a pretty positive thing for the family, when they get to take them away from the room. It's a step towards normality."

Carla Cucinatti, the unit social worker, has found that it takes about six months for patients to even begin to address their disfigurement. On Bigelow 13, no one views them as freaks, but once a burn survivor returns to the world outside, all that changes. "That's when patients begin to realize that being burned and having scars is for life. 'Now I'm living, now I'm doing the things I used to do, but I'm doing them as someone who's been burned. This is not a temporary situation; this is my life.'"

Confronting that was a huge challenge for Dan, one that took months, with many missteps along the way. There were times when he thought, "This sucks. I'm never going to be the same. I'm never going to be able to do this. I'm always going to look like this. It could have been very depressing, if I fed into, and at times I did." In time, he realized he had a choice. "I could sit there and get angry about it, get depressed about it, or do what I could about it. And for a while I did the former, got angry and upset, and then I was able to channel that into activity, and, you know, work on other parts of my person that weren't as directly related to my appearance. And to learn that I was a lovable person again. So I'm grateful that I didn't continue down the path of feeding into that depression or anxiety or anger."

It is never easy, but John Schulz believes, "The bottom line is the way the person deals with what's happened, disfigurement and all, has a lot to do with who they were before they got burned." People who have a strong will to live and a history of taking control of their lives, he has found, tend to take control of their recuperation and rehabilitation.

* * *

It was the last thing Peggy and Jack expected to hear when they walked into room 24 on March 19. Croaky and breathless though it was, there could be no doubt it was their son's voice. Dan was sitting up in bed, still attached to the respirator, but he was doing most of the work of breathing. Best of all, his tracheostomy tube had been upgraded to a fenestrated type, a major step toward normal breathing. By adjusting various parts of the tube assembly, Dan could breathe through his mouth and nose, cough and clear his throat and lungs of sputum, and—miracle of miracles—he could talk. Making the adjustments was complicated, so he needed help with that, and breathing, coughing, and talking were tiring, but Dan had been looking forward to surprising his parents and had spent the morning practicing, resting, then practicing again with the help of the speech-and-swallow specialist and the respiratory therapist.

After a roller coaster ride that had lasted nearly two months, the O'Sheas were smiling for the first time. Dan had been in room 24 for a little over two weeks, his condition changing on a daily basis. Sometimes for the better, often for the worse. One day Jack and Peggy would arrive to find him sitting up in bed deep in conversation with Mike Wilson, with Mike doing the talking and Dan listening and scribbling the occasional comment on a notepad next to the bed. Another time they would be met at the door with bad news: His gastric tube had pulled out of place and an abscess had developed in his abdominal wall; he would need surgery to remove the tube and he would be on IV feeds for a few days, then back to feeding through his nose. He had pancreatitis; it could be related to his burns or to his drinking, or both. Once they were told some of his skin grafts had not taken, so it was back to the OR for a few patches. Another time they found out the scar that was forming at his right underarm was contracting more than it should, despite rigorous range of motion work with the physical therapist, so it looked like he would have to return some months down the road for scar revision surgery. For Schulz and the other doctors, these were complications, yes, but not life-threatening ones, and they could all be dealt with. "From a professional standpoint," Schulz said, "every day wasn't a crisis. But it was just really hard for them, as parents. For them, it was an ongoing crisis. I'm sure it would have just felt like every day was a horrible day."

What was life-threatening, Schulz said, were Dan's repeated bouts of pneumonia, the most recent just a few days after he graduated to room 24. The culprit was not methicillin-resistant *S. aureus* this time, but it was still serious. The damage the smoke had done to his lungs made them vulnerable to every bug that was floating around, bugs that people with healthy lungs and intact immune systems can breathe in and expel with no more than a cough. "When somebody with severe smoke inhalation gets one of these pneumonias, they're kind of on a knife edge, and they can fall one way or the other," Schulz explained. That meant fighting each pneumonia aggressively with antibiotics and doing repeated bronchoscopies to clear Dan's respiratory tree of the casts and secretions that harbor infection.

The trouble was, they had begun gradually weaning him off the respirator, but when this new bout of pneumonia cropped up, Schulz told Peggy and Jack that they had to put weaning on hold, at least for a few days. More bad news. What did that mean? Another three, four weeks before they could take him home? Their spirits hit a new low.

Getting off a ventilator is not a matter of simply pulling out the breathing tube and turning off the machine. It has to be done gradually, over a period of time that can extend to weeks. Little by little, vent settings are turned down—decreasing pressure, volume, and oxygen content—while patients are watched carefully to see how fully they are able to inflate and then deflate their lungs, how regularly they inhale and exhale, how physically taxing it is, and how much oxygen is making it into their blood. The respiratory therapist might turn down the pressure so they are breathing along with the respirator for maybe an hour, maybe two the next day. "Our goal is always to move people back down to where they are breathing on their own, initiating their respiratory effort, with as little support as they need," John Schulz explained. Weaning off the powerful drugs that make being on the respirator bearable goes on at the same time, but none of it is straight line, and certainly not full speed ahead. Rather, it tends to be two steps forward, one step back, and not infrequently it is one step forward, two steps back.

By March 8, Dan's pneumonia was in check and he had his first "trach mask trial." For three hours, he was off the vent, breathing on his own while holding an oxygen mask in front of his tracheostomy. It was an exciting experiment, but also a risky one. The forceful push and pull of the respirator helps empty the lungs as well as fill them. The mask was just wafting oxygen into the trach (pronounced *tray-k*); Dan had to pull it in and push it out. If the lungs do not deflate completely, secretions can accumulate. Secretions are a breeding ground for infection. The cycle is vicious.

After two days of trials, Dan was wheezing a bit and a bronchoscopy found and sucked out more fluid. "It wasn't a real setback," Schulz recalled. "But every time we did a bronch, his parents thought it was." They turned up the pressure on the vent just a bit and stopped the trials for a few days, using regular bronching to keep the secretions under control. This stuttering improvement is par for the course, kind of like getting back into condition after a long time on the bench, Schulz said.

When they started up the trach mask trials again a week later, it was for shorter intervals, an hour at a time once or twice a day. That seemed to work better, and it was pretty clear sailing after that. No one wanted to tell Dan's parents about the fenestrated tracheostomy, which would make it possible for him to talk for the first time in months. For one thing, there was no guarantee that it would happen on schedule and no one wanted to raise their hopes. But mostly, Dan wanted to be the one to tell them, with his own voice.

When that day came, it was a wonderful surprise, and Dan's progress went pretty smoothly from then on. A week later, when his case came up in multidisciplinary rounds, he had been on the trach mask for thirty-six hours and was doing fine. That was the sign everyone was looking for: He no longer needed critical care. It was just a matter of downsizing the tracheostomy tube, transitioning to one that he could breathe around when he wanted to try it without the oxygen mask, and finally removing the tube entirely. The hole, covered with a Band-Aid, would gradually shrink closed, leaving no more than a dimple just below his Adam's apple to hint that the tube had ever been there.

When Peggy and Jack came in the next day, there was someone else in room 24 and their son was down the hall, a floor patient at last. Four days after that, on March 31, exactly two months after he had arrived on Bigelow 13, Dan O'Shea was discharged.

And then, he will be the first to tell you, the really hard times began.

CHAPTER TWELVE

New Skin

Every day the islands of epithelial cells grew larger, joining together like miniature continents in motion across the sea of granulation tissue. Soon both of Tom Parent's arms and hands took on a moist, mottled look, blotches of white and pink on a red background. All except for his left pinkie. Carl had been wrapping each finger separately, a painstaking task that kept them apart and movable. But for some reason that pinkie was not healing the way the rest of his hand was. Every time Carl unwrapped it, air hit the raw wound like a knife. Bending it at the knuckle was excruciating. Even when it was tightly bandaged and drenched with silver nitrate, the throbbing pain was enough to wake Tom several times each night. If it were not for that pinkie, his progress would be exemplary. After nearly three weeks, it was looking like he might soon be able to go home.

Tom and Nancy do not agree about who suggested it. Tom thinks it was Carl, Nancy says it was Dr. Ryan, but one morning there was talk about pigskin and the next thing Tom knew he had a piece of what looked like beige gauze wrapped around his pinkie. It was the actual skin of a once living pig that had been cleaned and processed into a meshed bandage. Tom was amazed. The pain stopped immediately and, for the first time since he arrived at the hospital, he got a solid night's sleep. "I owe it all to that pig."

* * *

Healers have been covering open wounds with animal skin since the beginning of recorded history. Frog skin was popular with the ancient Egyptians; along with lizard skin, it is still used in parts of the world

213

today. Human skin, from both the living and the dead, has been in the surgeon's kit for centuries. The problem, of course, is that skin belonging to another person or animal will be rejected, sooner or later, as surely as an alien organ will be unless the recipient's immune system is knocked out by strong antirejection drugs. It took awhile before anyone figured that out, and in the meantime some colorfully grotesque attempts were made. A Sicilian barber-surgeon of the sixteenth century named Antonio Branca is said to have made new noses with skin "borrowed" from his slave. In 1880, a Chicago surgeon named E. W. Lee grafted sheepskin to a young girl, who died before the graft was rejected, which made the exercise less than instructive. The skin of Mexican hairless dogs has also been tried. The only two mammalian skin replacements that continue to be widely used are pigskin (porcine xenograft) and cadaver skin (allograft). Neither, however, fills the bill as a permanent skin substitute. Rather, they are temporary wound coverings, their ultimate rejection by the human organism turned to advantage.

The ideal skin replacement comes, of course, from the patient's own body, but even autografts are not perfect. Scarring and pigment changes, poor healing, infections, permanently thin and fragile tissue, and significant pain are problems at both donor and recipient sites. The biggest problem, however, is one of supply and demand. Big burns are the ones in greatest need of grafting, but they leave little healthy skin suitable for use. Many big burns can be covered in stages, with an interval of a week or two to allow the donor site to heal, but there is a limit to how often a single site can be reharvested and a limit to how long a wound can remain uncovered.

One source of living replacement skin is a rare, but viable alternative to autologous grafts: tissue donated by an identical twin. Stanley Levenson told me a story of how he once was able to capitalize on this prime example of tissue compatibility. Sometime in the 1960s, when Levenson was director of the burn unit at Jacobi Medical Center, a patient was brought in with massive burns extending from his ankles to his buttocks. He had been burned while working with an acetylene torch; only his feet were spared, protected by his work shoes. Clearly, the man would need skin grafts, but it would be hard to find enough unburned skin on his body to cover what had been destroyed.

As a medical educator, Dr. Levenson always told his residents to ask if a burn patient had an identical twin. It would be a long shot, to be sure, but he maintained the question was worth asking. It was late at night when the senior resident called about the new patient. "He told me that he had asked and the answer was yes, there was an identical twin." The man's twin was willing to donate as much skin as needed. The patient had sustained severe inhalation injury as well, and his brother also offered to give up a lung. Fortunately a lung transplant was not necessary, but "we excised the burned skin totally and took the same amount from the brother's thighs, abdomen, and back, and used them to cover the wounds," Levenson told me. "The grafts took completely, and they stayed forever."

It is the rare burn patient lucky enough to have an identical twin. Nonetheless, John Schulz said, "You'd be amazed how frequently [unrelated] people ask if they can donate skin."

Ironically the need for replacement skin is a problem burn care professionals would not be facing if they had not done such a good job of keeping burn victims alive. In the past, people with big burns simply died, so not having enough healthy skin to graft was a moot point. With better survival rates, however, the quest for a permanent solution has taken on new urgency and taken researchers to the frontiers of polymer chemistry and tissue engineering. When it comes to artificial skin, the wow factor is huge, but the truth is that science is still a long way from giving nature a run for the money. Manufactured or tissue-engineered wound coverings may be analogous to skin in some ways, but the only permanent replacement skin that is structurally and functionally identical is a graft from the patient. A lot of heavy breathing accompanies the introduction of every new product that its manufacturer claims will be a viable substitute for human skin. The excitement, thus far, has not been justified by the result. Some "artificial skins" are useful in a narrow sense—as temporary placeholders until the body regrows its own. Others are based on interesting science but are impractical for actual use. A few suggest a promising new approach while others may, with refinements, someday fill the need. Those are a lot of "maybes" and "not yets." Tissue science is still far from its ultimate goal: a ready and abundant supply of new skin to replace what was destroyed by heat and flame.

A lot of different terms are used to describe these would-be tissue replacements: *artificial skin, human skin equivalent, dermal analog, skin substitute.* For the purposes of this discussion, they are referred to as skin substitutes—though it is in many ways a misleading term since they all lack blood vessels, hair follicles, sweat and oil glands, nerve fibers, and all the living cellular components of skin. As far as the current state of the art goes, the larger class of skin substitutes may be divided into two subclasses: biologic dressings and temporary wound coverings. Which is used depends largely on what was lost.

Superficial partial-thickness wounds that will heal without grafting need to be protected while they reepithelialize by something that stands in for epidermis as a moisture and microbial barrier as well as a means of relieving the pain that comes with exposure to air. The partial-thickness wound at the harvest site of donor skin falls into this category, as do scalds and other burns that destroy the epidermis and penetrate the dermis but do not wipe out all epidermal appendages. These burns have classically been covered by gauze dressings. Indeed, petrolatum-coated gauze, scarlet red, and its cousin, *tulle grasse* (literally "greasy net," a close-meshed gauze impregnated with vegetable oil, balsam of Peru, and paraffin), have long served the same purpose as biologic dressings, without the biology. More recently, a wide variety of specialized dressings ranging from polymer films to hydrocolloid gels have joined them.

Pigskin is the forerunner of manufactured biologic dressings. It is available as a fresh piece of split-thickness hide or as a meshed sheet of reconstituted homogenized dermis. In either form, it is no longer living tissue. As Tom Parent noticed right away, it lessens the pain of a raw wound. It also keeps the wound bed moist and protects it from trauma. Its principal benefit is that it adheres to the wound within twenty-four hours as porcine collagen forms a temporary bond with the wound bed, eliminating the need for painful dressing changes. When epithelium begins to grow beneath it, the pigskin gradually curls up and eventually falls off.

Biobrane follows the pigskin model. It is a silicone film bonded to nylon mesh, which in turn is coated with a gelatin derived from pig collagen. The silicone side faces up, providing an epidermis-like semipermeable barrier that keeps bacteria out while maintaining a moist wound bed. The dressing is applied with the nylon mesh side in contact with the wound, then stretched tight and stapled or taped in place. The gelatin coating acts as glue, picking up moisture from the wound over the course of a day or two and bonding to it. Unlike bandages made of flexible film with stickum around the edges, Biobrane is "rejected" as new epithelium grows beneath it, peeling away from areas ready to be uncovered. As the edges curl up, they can be snipped off.

Adding "bioactive" substances to a dressing supposedly improves healing by exposing the wound to growth factors and other substances associated with reepithelialization. That is the idea behind TransCyte, which can be thought of as Biobrane souped up with fibroblasts from neonatal foreskin. The product is cryopreserved (frozen), which kills the fibroblasts themselves, but the growth factors, collagen, and glycosaminoglycans (GAG) they produced remain. TransCyte may be used to protect superficial wounds while they reepithelialize and to cover deeper burns awaiting autografts, so it falls somewhere between a biologic dressing and a temporary wound covering.

Once they have been excised, deeper burns that require grafting must be protected until sufficient donor skin is available. If these burns are merely dressed, they may begin to heal before a graft is placed, but in the disordered fashion that produces abnormal scar tissue. Temporary wound coverings are said to form a scaffold onto which collagen will be laid down in an orderly arrangement. At the very least, what they do is provide a latticework into which granulation tissue grows. And, like granulation tissue, they eventually degrade. All are intended to be covered with a thin split-thickness graft—a slice of epidermis with a sliver of dermis—when donor sites are ready for reharvest and the wound bed is ready to receive it, two landmarks that it is hoped will be reached at the same time.

Allograft—human cadaver skin—is the only wound covering that by any stretch of the facts could be called a skin substitute. It is skin, but the skin of a stranger. Unlike porcine xenograft, it is alive (even though the donor is now dead), enabling it to engraft, albeit temporarily, as blood vessels from the underlying wound bed grow into it. It takes three to four weeks before the body notices the imposter, which is long enough to establish a foundation on which to place a thin autograft.

The principal purpose of allograft and the other skin substitutes is to buy time by reducing hypermetabolism, loss of fluids, infection, and pain. For example, allograft was the linchpin of early excision. The protocol John Burke established at Mass General in 1976 was excision of all dead tissue and autografting in stages; the waiting wounds were covered with allograft until donor sites healed enough to be reharvested. This was coupled with the BCNUs and silver nitrate to keep patients from succumbing to deadly infections.

An allograft, like a donor heart, kidney, or liver, is really an organ transplant, and along with blood transfusions, subject to the same regulations and risks, including contamination with viruses such as HIV and hepatitis. It is in limited supply and has complex storage requirements. It is kept in fresh or frozen form in tissue banks, which are accredited by the American Association of Tissue Banks and policed by the Food and Drug Administration. Clearly there is a need for something that is easier to come by and easier to keep. That something would have to be "manmade."

The first successful attempt at fabricating a skin substitute was Integra. A system designed to replace both layers of skin using a "dermal regeneration template," it consists of two layers that are structurally analogous to human skin. An outer silicone film temporarily serves the barrier function of epidermis. The inner layer is a matrix of collagen and GAG of animal origin that resembles the structural and molecular components of dermis closely enough to serve the purpose. The collagen guides the healing process by "showing" fibroblasts how to rebuild an orderly foundation of neodermis; over time, the matrix biodegrades at a rate controlled by the GAG. When the neodermis is ready to receive a graft, the temporary outer film is peeled off and replaced with a very thin autograft.

* * *

The story of Integra is best told by the man who invented it. For John Burke, finding a workable skin substitute was not an academic exercise or an isolated problem. Rather it was part of the larger campaign, as he put it, "to get into the modern world with burn care, by excising all the dead, right after the injury, and closing the wound immediately." Burke was among those responsible for the fact that patients with big burns were surviving, albeit with big open wounds in need of immediate closure.

The next challenge was to find a suitable material with which to close those wounds.

It was the late 1960s, the dawn, or perhaps midmorning, of the new age of plastics. "People were taking the newly developed technology coming out of the plastic industries, and every time a new material would come out of the research development labs—the pharmaceuticals, the universities, god knows where—someone would say, 'Won't that be an artificial skin?' So everybody tried to fit the output into artificial skin. When I began to think about it, I thought, that sounds backwards to me. It is very unlikely the plastics people, who are doing things having nothing to do with medicine or burns or anything else having to do with biology . . . they were just physical scientists wanting to make structural things. It would be about the same probability as a chimpanzee writing the Old Testament in the sand. You know, if he scribbled long enough he'd write it." Burke's narrative style is an intoxicating mixture of distinguished Harvard Medical School professor and cracker barrel philosopher. An iconoclast to the bone, he is also master of the mangled cliché. "So I thought, 'It seems to me what you ought to do is decide what the problem is and then try to engineer a material that would solve the problem.' Sort of going from the other end of the thing. And since I had my initial university training as a chemical engineer, I had some feel for the engineering—I hasten to point out that I had never done any engineering in my life.

"But at any rate, I began to think if you're going to replace skin, God or the evolutionary system or however you want to slice it, spent a couple of billion years figuring it out, did a remarkably good job, as a matter of fact. Skin works really well. So it seemed to be a little, if nothing else, *impolite* not to take some cues from the already established system. The major portions of the dermis, the two big classifications—outside of the cells, which I wasn't thinking about, I was just thinking about the matrix—the structure was collagen and glycosaminoglycans, what was called ground substance in those days." These days it is called extracellular matrix, or ECM.

"So I decided that if any engineering device had any possibility of working, we ought to start making it out of collagen and a glycosaminoglycan. And since I didn't know anything about glycosaminoglycans, I needed a glycosaminoglycan chemist. So I convinced David Swann, who had a big league reputation for glycosaminoglycan chemistry, to be part of this artificial skin effort."

The way Burke describes what followed sounds like a bunch of kids fooling around with kitchen chemistry. "For the next year or so we made some of the most unbelievable messes. Nothing looked even vaguely related to skin or dermis or whatever. So after about a year or year and a half of this, I said we better take stock here, count our losses, and see what is going on. And what I came up with was that I knew what artificial skin was supposed to be made of, but materials science wasn't my dish and I needed somebody that knew about materials to join the crew. When you do that at a university, you call all your friends. So I called all my friends at Harvard and said, 'This is my problem, do you know anybody who knows about this kind of thing?' And they all said, 'Well, that sounds like an interesting problem, probably somebody does, but I don't know anybody right at the minute.' So then you say, 'Well, if you happen to run across somebody, let me know.' And the next thing you do is you pick up the other phone and you call all your friends at MIT and tell them the same song and dance and they tell you the same thing."

Burke's networking finally paid off at the end of 1969. "As I remember, between Christmas and New Year's, the phone rang, and it was Yanni Yannas, in the MIT Department of Mechanical Engineering, and he said he was also interested in an artificial skin, and I said, 'You know, if you have any interest in this why don't we sit down and start talking.' If I'd have worked on this project by myself for fifty years I don't believe anything would have happened. And I believe if Yanni had worked by himself, nothing would have happened. But the ability to integrate the biologic problem with engineering solutions was the key to success." The road to success was a long one. Burke and Yannas were, after all, breaking new ground.

"We were not terribly bright, so it took us a long time to figure out how to do it," Burke told me, conjuring up the kitchen chemists once again. "I would guess, it was six or eight years. Yannas had his own laboratories in the Department of Mechanical Engineering aimed in one direction, as we did over here. And even though I was utterly convinced that I knew exactly what the material had to be—that is, it had to be collagen and a glycosaminoglycan—he was convinced, I think at the beginning anyway, that it ought to be a polymer. We tried the plastic material and that didn't work worth a nickel, and then we tried the collagen material and got, not success, but encouraging results, so we stayed with collagen. It's bovine collagen, comes from cow tendons now, but the original stuff we used came from cow hides."

The next step was clinical trials, years and years of clinical trials. Things were further delayed by changes in federal regulatory requirements. Integra falls under the authority of the Food and Drug Administration Center for Devices and Radiological Health, which came into being in 1990, in the midst of the Integra trials. CDRH finally approved it for use in burn patients in 1996, nearly thirty years after Burke's phone rang and his collaboration with Ioannis Yannas began. Five years later, in 2002, it gained approval for a second use in reconstructive surgery to repair disabling scars such as those that result from faulty healing of burn wounds.

Integra is far from perfect, as John Burke would probably be the first to acknowledge, but it paved the way for nearly every skin substitute that followed. It is best suited for flat surfaces, well away from any joint. The silicone film is inflexible, so movement causes it to shear off, pulling the GAG layer along with it. When Integra is applied to limbs, splinting and total immobilization are mandatory. Some surgeons find it hard to work with and some burn centers use it sparingly, if at all; others use it every chance they get and have developed strategies that raise its application to a high art. At the General, it is just one of many options, to be chosen when the conditions are right and the need justifies the expense and the risks weigh less than the potential benefits. As John Schulz said, "Integra is not a miracle. Sometimes it gets infected, sometimes it doesn't take. It's just another weapon in our arsenal."

There is no question that Integra has its uses. There is also no question that it is expensive and requires careful handling, including refrigerated storage until it is ready to be applied. There is an apocryphal but oft repeated story about an entire case that had to be discarded because someone failed to keep it on ice. At $1,000 a sheet, that is a lot of spoilage.

* * *

Around the time Burke and Yannas were working on Integra, another pair of researchers was exploring ways to expand the supply of replacement skin. Howard Green and James G. Rheinwald, who were in the cell biology department at MIT and later moved to Harvard Medical School, aimed their research at reproducing living cells. They focused on the stacks of keratinocytes that make up the epidermis and developed a method for growing keratinocytes in the lab, beginning with a small piece of healthy skin. They published the first paper on the process in 1975.

Their discovery harnessed the ability of keratinocytes to multiply by division as a means of amplifying available autologous donor tissue. As important as this was in theory, it has not turned out to be the thrilling answer to skin scarcity as it might appear at first blush. What has come to be called cultured keratinocytes or cultured epidermal autograft is just that: the gossamer thin outer layer of the skin. It has to be laid over some sort of dermal stand-in such as allograft or Integra.

The Green-Rheinwald process was licensed to BioSurface Technology, which introduced a commercial product called Epicel in 1987. It was eventually acquired by Genzyme, which is currently the only company that produces graftable cultured keratinocytes. Epicel is delivered topside down on a gauze carrier, making it possible to apply without tearing, though it has been described as having "all the strength of wet tissue paper." Not only does it lack the underlying structure and toughness lent by dermis, it is only ten to fifteen cells thick, so it does not provide the full five strata of normal epidermis. Once in place, it takes a month or more to thicken to something near normal. Until that happens, it is very delicate, prone to blistering, breakdown, and infection—if it takes at all.

Epicel is very expensive. Because it is custom-made from a snippet of the burn victim's healthy skin, and since no one plans to be severely burned, it cannot be ordered in advance and kept on hand until needed. It takes two to three weeks to grow a sheet the size of a playing card, and it must be used within twenty-four hours of leaving the lab.

As interesting as the technology is, Epicel is not widely used. According to Genzyme, the company currently fills orders for about 120 patients a year worldwide, and fewer than 1,000 patients have been treated with it since its introduction. Its significance lies in establishing the principles of tissue engineering as a potential source of living tissue. Everything that has made its way from lab to marketplace since is the conceptual offspring of Integra or Epicel or a combination of the two.

* * *

Some skin substitutes are built on bovine collagen; others rely on synthetic materials, and still others on a combination of natural and synthetic materials. Many add living cells or cell products, the principal source being neonatal foreskins, which is about as close as you can get to embryonic stem cells without starting an argument. One product that is still in development replaces Integra's silicone layer with autologous epithelial cells to make a sort of open-faced sandwich of Integra topped with cultured keratinocytes.

Regardless of the source, any time human cells are used in the manufacturing process they must be screened for a slew of common viruses and other pathogens. Many of the products are redundant, addressing a need already filled, either by pigskin and allograft or by the vast catalogue of low-tech wound dressings. All are expensive, as they must be to justify the huge investment in research and development. Many have come to the market courtesy of strategic alliances between small biotech companies, which did the initial R&D, and the pharmaceutical giants, which acquired the rights in return for their marketing muscle and a bundle of cash in the form of royalties. In some cases, the smaller company has gone belly up, in others the partnership has fallen apart and the original developer has reacquired the rights; it is often a murky business and in a few instances the supply of the product has been sporadic and unreliable.

All skin substitutes are regulated by the Food and Drug Administration, the majority of them under the auspices of the FDA Center for Devices and Radiological Health, according to the terms of the Safe Medical Devices Act of 1990. The development and marketing of medical devices is big business, and the approval process in the United States is lengthy and complicated. It also happens to be fairly easily manipulated. Once a device (or drug) has satisfactorily passed the three stages of clinical trials designed to demonstrate its safety and efficacy for the treatment or diagnosis of a specific disease or condition and has gained FDA approval to be labeled and marketed for that use, there is nothing to prevent its being used for a different disease or condition. The manufacturer is barred from promoting such "off label" use, though there are ways around that limitation. For example, a number of the skin substitutes were approved for the treatment of diabetic foot ulcers and venous legs ulcers, two types of nonhealing wounds that are far more common and thus have a market that dwarfs that of burns. A further loophole is provided by the humanitarian device exemption (HDE), which since 1996 has eased the approval process by requiring the demonstration of safety of a device that is intended for the treatment or diagnosis of a disease or condition affecting fewer than 4,000 patients but exempting it from clinical trials to demonstrate efficacy. It must only be shown to be of probable benefit in the absence of another existing approved product that provides an equal or greater benefit. Similar to the Orphan Drug Act, the HDE greases the skids by simplifying and streamlining a product's way to the market.

There is money and prestige in all this research for clinical investigators and the institutions that employ them. Although some research funding comes from public sources, most notably the National Institutes of Health, clinical trials are largely funded by industry. Lately some daylight has been penetrating the darker corners of corporate sponsorship of medical and scientific research, but a mutual back scratch continues to be part of the equation. This is not to suggest that the research does not advance the cause of medical science, nor that all or even most trials are biased, nor that drugs and devices that pass all hurdles do not benefit the patient public. It is intended, instead, as a caution to ignore the heavy breathing and tune out the hype. It all comes down to: Is this new product really better than the old standbys? The answer is a resounding "It depends."

It depends on the nature of the wound and what the surgeon is trying to accomplish. At the present time, there are two types of products that are more or less analogous to human skin, the less being those that act as a stand-in for dermal structure and the more being composites in which the "dermal" component is topped off with something that acts like epidermis.

There are two dermal analogs—one totally natural, the other a hybrid. Alloderm is cadaver skin that has been processed to remove all immunologically active components of the skin (the entire epidermal layer and all cells in the dermis) to prevent it from being rejected. What remains is the basement membrane and the extracellular matrix, but in this case the source is human rather than bovine or porcine. It is freeze-dried, then rehydrated just before use. Like unprocessed cadaver skin and products containing cells from neonatal foreskin, Alloderm must be screened for viruses and other pathogens. Dermagraft is a synthetic lattice made from polyglactin, a bioabsorbable polymer that is also used for the kind of surgical sutures that eventually dissolve. It is seeded with fibroblasts derived from neonatal foreskin. Dermagraft is frozen, which kills the fibroblasts, leaving behind human growth factors, cytokines, and GAG.

Apligraf and OrCel are sandwiches of natural and synthetic materials whose composite structures mimic skin, with analogs of both dermal and epidermal layers, separated by something resembling the basement membrane. Both use neonatal foreskin, supplying fibroblasts on a scaffolding made from bovine collagen topped with a semipermeable membrane on which keratinocytes, also from foreskin, have been cultured. Both products have to be used within days of manufacture since they are delivered fresh. Apligraf boasts a more lifelike epidermis because the top is air-dried to produce a "stratum corneum," the outermost cells that flake off the skin surface. Whether any of these differences actually makes a difference has yet to be proven.

It is worth a wager that legions of biomechanical engineers are toiling in laboratories around the world to develop skin substitutes that represent an improvement, or at least a patentable variation, on these themes. They may be new and improved hybrids of human cells and human-made materials or honest-to-goodness lab-grown human skin, cultured from biopsy or, where and when it is legal, stem cells. Still, the drum roll accompanying the introduction or even the rumor of a tissue-engineered solution to the scarcity of donor skin elicits skepticism rather than euphoria in the burn treatment community, despite the marketing efforts of the manufacturers, which manage to excite the media, which in turn raise hope in the hearts of burn patients and their families.

It takes time for new medical products to prove themselves, and prove themselves they must. Do they fill a need or are they just another option among many? Are the steep learning curve and expense justified by a clear superiority? In the end, will this particular innovative approach make a difference, right now, to this patient lying before me in the bed or on the operating table?

Someday someone may find the holy grail: a truly practical material or method for closing wounds too deep to heal spontaneously and too large to be fully grafted within days of the injury—something that looks and acts like skin, is readily available in quantity and perhaps even for a less than astronomical price, has a long shelf life and is easy to handle, carries no risk of disease transmission and does not create an environment that fosters infection, minimizes scarring and prevents formation of hypertrophic scars, and will not be unmasked as alien tissue and rejected by the body. In the meantime, John Schulz said, "We use whatever we can get, a combination of allograft, Integra, the patient's own skin, and amplified epithelium. Any port in a storm."

CHAPTER THIRTEEN

Burned to Death

When death comes to Bigelow 13, it may come suddenly or it may take days. It may come after an all-fronts battle or in the face of a "do not resuscitate" order. It may happen during the quiet of the night shift or the bustle of the day. It may come early on, during what is called the emergent period, before anyone has gotten to know the patient or family, or it may follow a long struggle and a deep attachment. Sometimes it is unexpected. Sometimes a decision is made to withdraw support, to let go of hope in a hopeless situation. Whenever it happens, it is felt throughout the burn unit—the pain, the grief, the waste of a young life, the sorrowful passing of an older one. For some it comes with a sense of failure, for others a sense of relief. There are always tears.

What do people die of when they die of burns? It could be cardiac or respiratory arrest, multiple organ failure, sepsis, pneumonia. They might bleed to death or suffer a stroke. They might die of a disease they came in with, their injuries tipping the balance. The burns may be massive, the enormity of the insult simply too much for the body to endure.

"Sometimes things just happen," Nancy Giese told me. "People are very sick. That's why they're in the intensive care unit. Sometimes you have an idea that an untoward event is going to happen, but a lot of times you don't. An aneurysm blows, somebody's lungs just fill up, they throw a blood clot, their kidneys just shut down, and it's the final straw on top of everything else. They can be fine and then they can be dead." She snapped her fingers. "It can happen that quick."

Many people who are burned never make it to the hospital. A jolt of high-voltage electricity from a power line or bolt of lightning can stop the heart as it sears its way through the body. An exploding gas line can kill with force as well as flames. A ruptured steam pipe can produce a wall of heat no one could survive. Smoke and noxious gases can kill long before an ambulance gets to the scene. People are trapped in burning buildings or incinerated in fiery crashes. Burn victims may die in the ambulance as it speeds toward the hospital or get as far as the Emergency Department but never make it to the burn unit.

In hospitals, especially hospitals on television, lives are saved and lost in the high-intensity atmosphere of a code blue. That rarely happens on Bigelow 13. The way Bob Droste explained it, most of what goes on during a code is what they do as a matter of course for their critically burned patients.

"We have very few code calls here. If I have a patient with a central line, an A-line, and he's intubated, we don't need the code team. They're just in the way. Quite literally. There's too many of them and there's too many bosses, so I don't want to hear it. If I have a question, I'll ask the burn resident: What do I do? and he'll tell me."

If a critical patient takes a sudden turn for the worse and a code is called, the procedures and protocols are very clear, making for what Nancy calls organized chaos. "The staff up here for the most part is very experienced, they have worked here for many years, so it's not their first time going through this," Nancy said. "Everyone in the unit is hooked up to a monitor, so if somebody codes, you will hear it. You hear something that makes you look at the monitor and then . . ." Nancy mimed a freaked-out Help! "If somebody in the unit says they need help, people drop what they're doing and they all come running. It's a team, and they will do whatever it takes.

"You watch your airway, your breathing, and then your circulatory system. So whatever part isn't working, the available staff will try to do whatever it needs, whether it means AMBU-ing, putting a mask on the person; if they're intubated but there's no air passing, they'll try to get air in by a mechanical method. If their heart has stopped, they'll do CPR, to be that artificial pump, to get that heart going. Somebody will go down the hall and grab the defibrillator, if the heart needs to be jump-started."

In the meantime, somebody has called the burn doctors and, if necessary, called in the code. "Our guys might be here or they might be in the operating room, so the code team may arrive before our service," Nancy said. By the time anyone arrives, the crash cart will be at the bedside, open and ready for action. The cart is a universal piece of equipment throughout the hospital. Nancy described it as a metal toolbox on wheels, its drawers filled with all manner of life-saving equipment. "The top two drawers have medications, the next one has respiratory equipment, the next one has pacing things in it." Everything that goes on during a code has to be thoroughly documented—it is a legal necessity—so a designated member of the team stands by and writes down everything that happens, all drugs and equipment that are used, regardless of the outcome.

Most of the time, the scenario is not so frantic, though it can be equally intense. "Most of our patients don't do anything sudden, so you see it happening, and as it happens we start to make changes," Bob said. "Sometimes we're giving them maximum support and it isn't enough. You don't call the code if you're already doing everything you can. Sometimes when you're doing all of those things over a period of time it's called a chemical code. And we've run a chemical code for days. It's essentially what you do during a code call, only we're doing it for days at a time. On TV they say: 'Get the epi!' Well, we run epi drips for extended periods. We've had people on pressors—cardiac support drugs—for weeks and weeks and months even. And they can be very unstable, inching back a little, forward a little, for a very long time. I'm not kidding that it can go on for days. You just change shifts and you pass it on." More than one nurse has gone home not knowing if a patient will still be there the next day. No one likes it, but it comes with the territory.

"Sometimes it fails, and sometimes they recover, even though it looks as if there's no chance. Occasionally we have somebody we're wrong about and they do get back, especially if they're young, and then you just flog them and you don't quit."

When someone on the unit is crashing, it is all hands on deck and a lot of long nights for the attendings and the residents. The doctors may direct much of the decisionmaking, but the burden of the work falls on the nurses, who must care for both patients and families. Bob remembered one recent case. "We had a fresh burn that came in, I think he was close to 80 [percent TBSA], and it took two nurses to take care of him for days, and the rest of us had to help those two just to keep up. We had a machine to pump blood in that could pump in a unit in about one minute. One day I pumped in ten in ten minutes, because he was just exsanguinating, bleeding out."

"We fight tooth and nail, right down to the line sometimes and they go anyway. And sometimes we withdraw care, if it's a lost cause. We don't get to those easily; many places do that way before we do," Bob said. If the patient is very old, the team is less likely to resort to what Bob calls "flogging"—aggressive life-saving measures. It may make things worse, subjecting a frail body to more stress than it can take. It may go against the wishes of the patient or family. It may, in the experienced view of the burn team, be clearly futile.

* * *

Futility. Withdrawing care. Comfort measures only. Do not resuscitate. These are loaded words that inhabit every intensive care unit and haunt the corridors of every hospital. There is no medical meeting in any specialty that does not include a session on the subject. In the burn care community, it is a side effect of the progress made in the past twenty-five years. It is now possible to save the lives of critically burned patients who previously would have died. Mike Wilson voices the dilemma quite succinctly: "Looking at the history of burns, it's tremendous how far we've come. All over the world people die of 20 and 30 percent burns. We can keep people alive here with 90 and 95 percent burns. The key is that we don't often enough ask the question 'Should we?' as opposed to 'Can we?'"

Twenty-five years ago, that question was asked at a meeting of a Consensus Development Conference on Supportive Therapy in Burn Care. Bruce E. Zawacki, MD, who spent a career thinking and writing about the ethical dimensions of the futility debate, asked, "When should maximal therapy be given to a burned person with injuries so severe that survival in similar cases has been rare or unprecedented?" In his view, maximal therapy is imperative unless the patient is competent and refuses it, in which case the patient's wishes must be followed. The consensus held that with burns exceeding 70 percent TBSA, physicians have a "moral responsibility to initiate therapy for all patients. Physical and/or emotional shock in the burned patient make it impossible for the victim to contribute to the early decisionmaking process."

Zawacki may be the most articulate voice on the subject but he is not alone in acknowledging that doctors often have difficulty engaging in dialogue with their patients over issues of life and death. They often do not know what the patient, family, nurses, and others know, and they are uncomfortable with the process of finding out. Indeed, it is a process, not a single conversation.

Mike Wilson insists that the only ethical way to make end-of-life decisions is to take as much time as possible and hear from many quarters. "I believe that in a process, a group process with the family who knows that person, the best medical advice we can get, ministers or spiritual advisers, that kind of thing, that with a collective process—not just one meeting—we'll be led to the right decision. I believe we have to enter that process, and I don't believe we always enter that process. There's a human journey involved in that and we're not educated to facilitate that journey. Doctors aren't trained to do it, and they don't do it; nurses aren't trained to do it, and they don't do it." For Mike it is a crusade, to which he brings not only his training as a clinical pastoral counselor, but also his willingness to speak his mind. "In other countries it would be considered an insult to wait until someone is dying to ask what their wishes are. Leave it to the United States to take the communication out of death. Until you die."

In the relatively rare instances when a critically burned patient is able to express his or her wishes, or has done so in advance and hospital personnel know what those wishes are, knowing how far to go is clear. More often, the decision is up to the patient's family, if a family can be located, and even then, family members do not always agree. Part of the work of the burn team involves working with families to come to the difficult decision when the news is bad.

When someone arrives in devastating condition, John Schulz told me, "We do our best to immediately start a conversation with the family, and try to present a realistic picture of the prognosis, and at least start a discussion of whether we should be treating someone." Nonetheless, he said, "Even if it is clearly futile, we are bound by the proxy's wishes. But we will continue to say what we think."

There are times when family and burn team are on the same page and, as painful as the situation is, the conclusion is foregone. A young woman was brought in with a 50 percent burn, deep third- and fourth-degree from the waist up, charred bones visible where skin, fat, and muscle had been burned away. Her arms were so thoroughly cooked, they seemed mummified. It was obvious at least one, probably both, would have to be amputated. While she was still downstairs in the ED, they had sliced through the eschar that armored her chest. Now the flesh was falling off in chunks. They were pumping fluid into her as fast as they could, and it was gushing out of her in equal volume.

What kind of fire could wreak such havoc? A fire the woman set herself, convinced that someone was trying to cremate her. She had been in and out of psychiatric hospitals for years, her anguished mother said. She had set herself on fire at home and it was her father who called 911. Now the parents were sitting in the conference room at the back of the floor trying to explain the unexplainable. Their daughter had stopped taking her antipsychotic medication. They had tried to make her take it, had tried to get her back into the last hospital she had been in, had begged for help everywhere, all in vain. They had fought for a life their daughter no longer wanted. Now it was time to let her get her way. When her blood pressure plummeted and her heart stopped, everyone knew that the family wanted to let her go. This was not the first suicide attempt she had made, but it would be the last.

At other times, what is very clear to the burn team is hard for a family to accept. "If somebody comes in and they're absolutely charred everywhere, they've got no chance, and everybody looks and says, 'Right,' then you do comfort measures from the beginning. And that happens occasionally," Bob told me. "It's not frequent, but it happens. Many times it takes awhile to go through everything, and get the family in and get the family involved." In general and if there is time, the person who has made the strongest connection with the patient's family initiates the discussion. It may be one of the doctors; often it is a nurse. Social work and psychiatry may get involved, especially if there is disagreement among family members or a gulf between what the family wishes and the burn team believes is possible. If that gulf is unbridgeable, the hospital's Committee on Optimum Care will step in. This is the MGH ethics committee, established in 1973 to address and mediate end-of-life questions. Volunteer members from many departments serve on the committee. Not surprisingly, Mike Wilson is one of them. "We haven't had to bring a case to them in a long time," Mike said, though "we may be coming up to one right now."

It was Mike's patient, a man with a 95 percent burn and, even more serious, significant brain injury. A CT scan showed herniation, the result of oxygen deprivation. "He breathed so much smoke that part of his brain is not functioning anymore; it's totally swollen. We never know, when the swelling goes down, what's going to be left. He could be left a vegetable or he could be someplace in between, but the combination of that and a 95 percent burn—thank God—makes us say not 'Can we keep him alive,' because we can. We're finally starting to ask the question, 'Should we?'"

The problem was that the family was not ready to let go. There were cultural and religious issues, as well as the fact that others in the family had been injured in the same fire and the situation was simply overwhelming for those who were praying and waiting. Mike and other members of the team had been meeting with the family since the man came in four days earlier, trying to help them understand the medical implications. The CT scan was part of the process. "If the medical evidence says going forward from here is futile, but the family isn't really ready to let go, then this would be a case that may go to the ethics committee, because we have to enforce the idea that we are doing him a human disservice" by continuing care, Mike explained. A phone call summons a member of the committee, who comes up, reviews the chart, and makes a recommendation. "All that does is if the family still says 'I want my loved one to live,' it starts the process of being able to legally withdraw care, because of futility."

The tricky part, of course, is that it is not always possible to know for certain who will live and who will die. There are patients on each end of the spectrum whose prognosis can be accurately assessed, but there are other patients whose course is unpredictable. The team can bring the best medical knowledge to the question and still be wrong. Mary-Liz Bilodeau remembers two patients who came in at the same time, from the same workplace accident, a horrendous explosion that left them both gravely injured. One died the first night. The other had burns over 80 percent of his body and was not expected to survive. "We could just as easily have said he was futile, but that didn't happen." That patient walked out of the burn unit. Not soon certainly, but the man lived.

It helps to be young and it helps to be healthy. In the past doctors used this rule of thumb: TBSA + age = risk of dying. For example, a twenty-five-year-old with a 50 percent burn had a 75 percent chance of dying; the risk would be the same for a sixty-five-year-old with a 10 percent burn. Survival rates have improved with improved treatment, but it remains true that a small burn a younger person could easily survive may kill an older person. Colleen Ryan was lead author of an article published in the *New England Journal of Medicine* that proposed a formula for estimating the probability of death from burns. Age over sixty, a burn larger than 40 percent TBSA, and inhalation injury were identified as the three major risk factors for death. Dr. Ryan and her colleagues reviewed the outcome of 1,665 patients at both Mass General and the Boston Shriners hospitals from 1990 to 1994, and concluded that a burn patient with none of those risk factors has a 0.3 percent chance of dying; one risk factor brings the likelihood to 3 percent; two factors increases it to 33 percent; and all three factors shoot the risk of death up to 90 percent.

* * *

Everyone on Bigelow 13 works hard to speak clearly, realistically, in ways families can grasp. It is not easy. "We have to comfort the families before the event has occurred, to let them know that it's okay, everything has been done that can be done, they're not killing the patient by withdrawing support or by saying no extraordinary measures. Which is just a hard thing," Nancy Giese acknowledged.

It is difficult to overstate the importance of communication. When a patient is in critical condition, the situation may change from day to day, which means difficult conversations may have to take place more than once, but what is said and what is heard are not always the same thing. The O'Sheas are a case in point. The seesawing course of Dan's illness was torture for Jack and Peggy. They clung to every hopeful sign and sometimes had difficulty hearing bad news. When John Schulz spoke to them down in the Emergency Department the first day, he made it as clear as he possibly could that Dan's survival could not be guaranteed. Nor could he predict the effect the carbon monoxide poisoning would have on Dan's mental status if he did recover. Jack and Peggy heard what Schulz said, but they focused on the subtext: This doctor is going to make sure our son lives. Later, when every day seemed to bring a new setback, it was harder for them to feel hopeful. Hearing, "You have to understand that your son is very very critically ill," Peggy's response was, "Oh my God, they're telling us he just isn't going to make it." The message was pretty much the same in both cases; the difference was the filter through which it was heard.

From the comfortable distance of four years, not to mention the certainty that he did survive, Dan applied his own filter: "I know honestly they had a sit-down talk at one point, I think with Dr. Schulz, trying to figure out how I was going to do. And he was up front in saying, you know, there was a fifty-fifty chance that I may or may not live, because the smoke inhalation had been so bad. My body was so beaten up that it was just touch and go there for a while, and I was lucky that I was able to hold on."

Doctors have a reputation for not wanting to let go. They consider it a defeat. They take it personally. Their training is to save lives, not to ease the passage at the end of life. When a junior resident and John Schulz were discussing a particularly difficult case—a patient who had already had one leg amputated and could well lose a foot on the other one, and a family in such deep denial that they expected their son to someday walk out of the hospital—the resident said, "It's so sad." Schulz replied, "We're here to make it unsad."

Still, Mary-Liz believes there has been a change between the last generation of doctors and the current one. "Certainly we wouldn't be where we are if it wasn't for the people that were treated and treated and treated." Nonetheless, she said, there was in the past a lot of unhappiness among the nursing staff about the sense that they were inflicting pain on patients just to keep them alive. "You know, as nurses at the bedside, you feel as though you're torturing someone and for what? We now have a younger group of surgeons who are willing to say, 'We can't save everybody and we have to look at this in a different way.' And I think the dialogue has been opened immensely. Are we perfect? No. Are we still at times maybe not making the best decisions or not entering into those discussions well enough, soon enough, often enough? Absolutely."

Looking at the futility issue is like looking through a prism; what you see depends on the angle of view. One nurse told me how upsetting it was to hear a social worker refer to a severely burned patient who had been hospitalized for a very long time and would undoubtedly be profoundly disabled as a "bad save." She hastened to add that the person who said it was from a different department. "We don't think that way. No one up here would say anything like that," she insisted.

* * *

Death does not pervade Bigelow 13, but it takes its place there alongside pain, alongside healing. Once a decision has been made to give comfort measures only, nurses spend a lot of time with the family, to whom the comfort measures must extend. As Nancy Giese described it, a lot of the care at that point goes into "making sure they're okay with what's going on and answering questions. Because this is the last time they're going to see or be part of this person's life. As well as just the regular physical work that you do with every patient that comes through, it's a lot of emotional and spiritual and being there, just being there. Some people are adrenaline junkies, which is great, you need those. Some people like to do the touchy-feely, cry with the family, which is good because you need to do that too."

She says she's one of the adrenaline junkies, but she also confesses to being someone who cries. "I'm one of these people, I cry all the time. That's what I do. I don't blubber, but I'll just tear, and a long time ago I stopped trying to even . . . I'm going to cry, it's part of me, it's who I am. I'll walk away so I won't upset anybody, but that's life, that's what happens."

Others may not cry as readily or unapologetically as Nancy, but everyone is affected when a patient dies, even the families of other patients. During Dan's stay, there was more than one death on the unit. Those deaths shook Peggy and Jack, and have stayed with them. Jack thinks that at least half the patients who shared the unit with Dan never made it home. They can list them: the man whose entire family came together to hold a daily prayer service at his bedside, the old woman whose adult children fought about whether to pull the plug until her death took the decision out of their hands, the young man who was transferred from another hospital even though it was really too late. He was just about Dan's age, except he was married and his wife was there the whole time. It was really heartbreaking. "One day you would go in and there would be an empty bed, and you just knew," Jack said. "I think one poor guy came and went overnight. After a while you didn't ask anymore. It's devastating to the nurses when they lose a patient, you could tell."

It is hard to know how the doctors feel. John Schulz, the man who is happy to explain anything, answer any question, who relishes any opportunity to teach, is uncharacteristically reticent when it comes to talking about death. "We're not easy to demoralize. You have to be able to reflect and kind of get in touch with that inner person." With a touch of irony intended to deflect any hint of emotion he might have betrayed, he added: "Since we're all surgeons, we don't believe in getting in touch with our inner persons." Unit psychiatrist John Findley disagrees with that assessment: "Fortunately the surgeons here are not that sort of classic malignant, hard-assed surgeons. They're the softer, kinder, gentler type."

Later, Schulz acknowledged as much. "It's very emotionally taxing. And when we have tragic situations in the unit, a lot of times it takes awhile to recover. It certainly takes me awhile to recover. But we have enough success that it kind of balances out. Personally, I feel that if we're doing our best, tragic situation or no, if we're doing our best . . ." He could not finish the thought. "It takes time to process emotionally trying situations. I'm fortunate to have two other people doing this with me so we can share that burden. But I don't have a good answer for that. The emotional intensity varies, depending on the circumstances of the burn, the age of the patient. To have an eighty-year-old who gets burned who led a decent life, that's a whole different picture than a twenty-six-year-old that's badly burned. Or a child. That is just . . . I can tell you, that just knocks our socks off and it takes weeks to get over."

CHAPTER FOURTEEN

A New Life

Dan O'Shea sat on a bench in Back Bay Station, his over-stuffed gym bag at his feet. His father had put him on the commuter train into the city, had said good-bye in a way that was more sad than angry, and here he was, feeling about as low as you can feel and still be breathing. And he was hung over.

The past six months had not turned out the way he expected when they checked him out of the burn unit, not that he really knew what to expect. But whatever it was, it did not include those three exhausting weeks at Spaulding, the rehab hospital—a kind of hard work he had never experienced before in his life. On Bigelow 13 they were gentle and seemed to understand what he was going through. But at Spaulding, the therapists were tough and it was serious business, all the time. It did not help one bit that his roommate moaned all night long so it felt like he never got a decent night's sleep.

Dan had looked forward to going home. Well, home to his parents' house. There was no way he could go back to his own place. His father had had it cleaned out and everything went into the dumpster. What was not burned to a crisp was waterlogged or reeked so badly of smoke that there was no point in trying to save it. So he pretty much had to start from scratch, and part of the trouble was he did not know how to do that. Going home to his parents' house ended up being no picnic. For one thing, he was a thirty-year-old man going back to live with his mother and father and hassling all that stuff they never could see eye to eye about. He knew they had gone through a lot too, worrying about him and everything, but it felt like everyone was on his case.

Every day the visiting nurse would come to change his dressings and check his breathing, his lungs and all. But it was nothing like the nurses at the burn unit, who really seemed to care how he was feeling. These nurses were strictly business, and half the time he did not even know their names. Next the therapists would come. Two or three hours of back-to-back physical and occupational therapy. It was exhausting, and all he really wanted to do was sleep, but everyone kept saying he had to eat more and do the exercises and put little pegs in holes like he was back in nursery school and practice cutting food with a knife, opening a carton of milk, pouring a glass of juice, stupid stuff like that.

Back when he was still on Bigelow 13 he had had to learn how to walk again, dragging all the tubes and hoses along behind him as walked from the bed to the door. That was a big deal. Just getting his balance back took time after doing nothing but lying in bed for weeks. As soon as he was steady on his feet, it was out into the hall, up and back. A few days before he was discharged he had managed to make the entire loop, from his room all the way down to the nurse's station and up the other side. Everyone applauded as he rounded the corner, and silly as it seems, it felt good when they did that.

Going back for clinic every week felt good too. It was about the only thing that did. He liked seeing Mike and the other nurses, talking to Dr. Schulz. They said his scars were looking okay, though Dan did not think so. They were hideous, he thought, and they still hurt. A lot. But at least the people there did not look at him like he was a freak, staring at his scars or turning away. And he was not the only one there, at clinic, who looked like he belonged in a sideshow. Not that he got into conversations with anyone, but at least he did not feel out of place the way he did everywhere else.

He was still angry, though, about the scar release surgery. The graft at the top of his right arm had not healed quite right, or something like that. They called it a contracture and said it kept him from having full range of motion. No matter how much stretching he did with the physical therapist, that arm was not the way it was supposed to be. So in August they said to come back and have it fixed. For some reason, it sounded like a casual thing: "Just come on back and we'll do a release." An alarm bell should have gone off in his head when they said he would be in the hospital for a week. They said they would cut out the scar and do another graft, and he had no idea, none at all, how painful that was going to be. After all, he had been totally out of it when they did it the first time, and by the time he woke up, the place on his back where they took the skin from was pretty much healed. This time, they took skin from his thigh and he woke up the same day. There he was in a room out in the hall, far away from where he was used to being, with a pump so he could give himself morphine whenever he had pain. And boy, did he have pain!

The funny thing was—well, not funny, but strange—what hurt was not his shoulder, where they cut deep through the scar and sewed on the graft, but his leg, where they had shaved off a thin bit of skin. He just was not prepared for how much it hurt. It was okay if he did not move, but morphine or not, the minute he got up to go to the bathroom or something, or even turned in bed, the pain was excruciating.

It might have been worth it if the new scar was any better than the old one. But it was just as bad, maybe worse, even though he had had a big argument with his father about that. His father insisted he needed to have it fixed, if he did not do it, his arm would be frozen halfway out from his body. But Dan was not sure. And it just got him really upset, after all he had been through, to have this new thing to deal with too.

That is probably one of the reasons he started drinking again. Just as soon as he could get out of the house and off his parents' leash, he hooked up with some of his old friends and they would go out, have some beers like they used to. Dan kept himself covered up, long sleeves, long pants, a Red Sox cap, and no one really asked him about what had happened. They knew, but they also knew he did not want to talk about it. He would come back late, try to get upstairs and into his room without waking his parents, but he knew his mother at least always heard him. And she knew. She did not say anything, but she knew.

Peggy remembered one night that was just the last straw. "He came in and he was just so drunk he could barely stand up." She heard him crashing against the wall trying to get upstairs. The thing that bothered her the most was she knew Dan figured he had a right, after all he had been through, to cut loose a bit. He had been really angry since he got out of the hospital, angry and self-pitying. "You know, all that 'Why did this happen to me?' stuff. And of course, you're ready to murder him." After they had prayed so hard for him to stay alive, now he was wrecking his life. As Jack said, "He still just wasn't getting it." They both knew this could not go on.

"So the next day, it was my brother's anniversary mass, and I said to Jack, 'I'm going out, and you're going to have to tell Daniel. You're going to have to tell him . . . and I'm not going to do it.'" It was four years later, but as she told the story, the whole situation suddenly seemed as raw as it had been during that terrible time. "He just has to know if he doesn't stop, he has to leave, that's all. You have to tell him, because I'm not going to be the one.'"

Peggy went to church and afterward to a park up on a hill where she liked to watch the sunset, and just sat there and cried, while Jack drove their son to the station.

Peggy still cries when she talks about it. "We did the right thing, I know it now, but we didn't know it then. But you really have to face them with reality: 'You can't go on using the burns and your life and all that to excuse your alcoholism. You can't do it.' It was not good for him and it was not good for us. And God, it's hard, but you just have to say it."

So there he was, sitting in the station: "homeless, jobless, friendless" is the way he remembers it. Looking back on it, he realizes that at a certain point his recovery from his burns had hit a brick wall. "I was home at my parents' trying to deal with what I perceived as the new reality and I was not dealing well with it. I was so frightened or angry or upset at the whole thing, and the whole recovery from burns is not an easy one, and so there were times I thought that I would never recover, that I was going to be a scarred person and *less than* for probably the rest of my life because of this. I felt as if when I came back into the real world—interacting with people outside of the hospital, in a work environment, meeting new people socially—the first thing I had to do was to explain my scars. I thought the scars that I had were going to affect every aspect of my life."

It was only about six months after he had been discharged, not much more than a month after his revision surgery, and his scars were still changing. It was difficult to believe, even though he had been told more than once, that the red and swollen evidence of what happened to him would someday fade.

And it has. Four years later, there is a shadow of a mesh pattern on his arms, the edges slightly hypertrophic, the whole thing looking sort of smeared and oddly hairless, but the color is the same as the rest of his skin and from a distance the scars are barely visible. The skin on his hands is unnaturally smooth, like he is wearing gloves. One finger is crooked a bit from a contracted scar, but when they offered to fix that with another surgery, he said, "No, thank you. I'll live with it."

"A lot of my anger was directed at myself because I felt like I did this to myself. I had had struggles with alcoholism before this whole incident and this to me was an extension from the drinking. I started to drink again and the results of my drinking were so dismal, you know, that it became almost a life-and-death choice for me. Am I going to continue life in this fashion or am I going to look for help?"

In a strange way, it is easier for Dan to talk about his drinking than about the fire and his burns. He refers to the fire as "the incident," "the accident," or "what happened." He has no trouble calling himself an alcoholic. Perhaps that is because of where he turned for help.

"I was aware of Alcoholics Anonymous. My father had been sober through that program for many years and I had flirted in and out with it myself for a couple of years. Ultimately, I was at the point where I was so low that I had to make a decision that day whether to throw in the towel or give it one last shot."

Dan O'Shea marks two major dates in his life: January 31, when he was saved from the fire, and October 23, when he saved his own life. "It was 7:30 in the morning and I was in the basement of Trinity Church and I raised my hand and cried out loud and asked for help. I think I cried really for the first time since everything had occurred to me. I mean, I was angry, I was upset, and I was in pain, but through the whole recovery process I never really wore those emotions on my sleeve. It was kind of an epiphany for me to be able to raise my hand and ask for help and then get it. Spiritually, that's the day that the grace of God entered my life. It was kind of like a wretched soul being saved. Whenever I hear that song, 'Amazing Grace,' I think of that day, always."

When he talks about it now, Dan will say that he felt he owed it to all the people who had worked so hard to save his life not to throw that life away: "From the firemen who came in and got me out of the fire, to the people at Mass General, the doctors, the nurses, my family." Especially his family. "My parents were there to support me every day at the hospital, when I wasn't conscious, when I was conscious, I mean every single day they would show up. And then they had to watch me destroy myself again. As much as I had gone through physically, I was still running myself into that rock, beating myself up, taking punishment through drinking. It was an emotional bottom that I reached, and then I was able to turn things around."

It was no more a straight line than his physical recovery had been, and there were plenty of bad days, but Dan was taking it one day at a time. "That's how things worked for me, miracles worked that way. You know, just one day moving into the next day, physically getting back on track. Moving out of my parents' home, going to the gym on a daily basis, and getting involved in activities I didn't think I'd be able to do again."

There came a point, Dan says, "when I sort of took ownership of my own therapy. I could drive now, I could get myself to a gym, and move beyond the stretching exercises and the sort of simple motion exercises that they had me doing. To me that wasn't enough." Dan traded the treadmill for the Stairmaster, and started lifting weights. "I was determined to get back to good health. I guess it was all going to happen anyway, but I was just pushing it. I refused to be limited from physically doing what I thought I could do before."

Getting back to the gym helped Dan overcome another major obstacle. He was in Boston, living with an old lover, who let him stay as long as he stayed sober, and he had a job. "I was getting up at 5:30 in the morning to be at the gym at 6:00 so I could be showered by 7:30 to be at work at 8:00. Get off at 8:00, go to a meeting at night, and then go home and get ready for the next day. And I was doing that every day. If I wanted to go to a meeting every night, the only time I had during the day to go to the gym was in the morning, on the way to work. I joined back up at the gym that I used to go to in the South End, and you know, there are a lot of gay guys who work out there and there's a lot that has to do with physical appearance and stuff like that. So if I'm going to go to work from there, I have to shower at the gym and that was a big traumatic choice for me: Am I going to disrobe and, as bad as these scars are and as bad as I feel about them, am I ready to do this? And I just had to say 'Yes, it's the only time I can do it, I can't give up anything else, and screw 'em, you know, if they can't handle it, then it's their problem, not mine.'"

Amazingly, the man who spent two months on a respirator, whose lungs nearly gave out on him more than once, now plays league basketball in the winter and softball in the summer, and takes a run around

Castle Island in Boston Harbor every day after work. That is something no one at Mass General would have predicted—not John Schulz,
not Mike Wilson, not the physical and respiratory therapists who
worked with him, and sometimes against him, for months on end. And
it is amazing that he does it in a tank top and shorts that do nothing to
hide his scars. That is something his parents would never have thought
possible.

* * *

They might not have let another patient go home. Tom Parent's arms
were still covered with dressings that needed to be changed every day.
The wounds were not fully closed and infection was still a possibility, so
a topical antimicrobial was a necessity. They had been dousing him with
silver nitrate, a regimen that cannot be followed outside the walls of
Bigelow 13. If it were not for Nancy and Acticoat, he would be looking
at another week or two in the hospital. But he was getting antsy to get
out of there, and Nancy was pretty exhausted trying to fit a half day of
work and being home for Ashley with a daily drive halfway across the
state so she could spend time with Tom. After a serious sit-down with the
doctors and intensive lessons from Carl, it was decided that Nancy could
do the wound care. "Dr. Ryan and Dr. Sheridan said to me, 'Look, we
really feel you can handle this. Would you feel confident handling this?'
And I said, 'Yes, I would.'" So in the middle of June they sent Tom home
with a carton of Acticoat and instructions to come back once a week for
clinic and to call if there were any problems in between.

With Acticoat, Tom could get his silver without the stain. It was just a
matter of laying the pads on his wounds, wrapping them in gauze, and
keeping it all wet with sterile water. "It's kind of like the home version of
silver nitrate," Tom said. Nancy would soak the dressings before Tom
went to sleep and then wake up once in the night to do it again. When
they drove in to Boston for clinic on Thursdays, they would break the
trip at the halfway point to douse it again, and so Tom could take his
pain medication. The alternative was silver sulfadiazine, which Tom was
not enthusiastic about. "Because when you change the dressings, you
have to get all the old stuff off." He would never forget the agony of having his wounds scraped.

At first they had a visiting nurse come do the dressing changes, but it turned out Nancy knew more about burn care than the nurse did. "You know, people don't know how to deal with burns like this. And Baystate Medical, which is our big hospital, they have four burn beds, so that's how many burns they deal with. They won't deal with burns that are more than 30 percent of the body. Well, if this had happened in western Mass, he'd probably have been stuck there and probably would have gotten much worse, because community hospitals just don't know how to deal with that. The visiting nurse we had didn't have a clue, literally. The first time she came in she wasn't going to wear a mask, and I said, 'Don't touch him!' And I got Mass General Hospital on the phone and said, 'You have to talk to them before you do this.' And they told her: 'You need to wear a mask. You need to be very careful with an open wound.' And we actually were the ones who showed her how to do the wrappings."

Tom was still on strong pain medication, Oxycontin, when he left Mass General, and there was no question he still needed it. He and Nancy were supposed to go to Ireland at the end of July, and he had no intention of taking a bottle of the notorious narcotic through Customs, even with a doctor's prescription. They had been planning the trip for a long time, since well before Tom was burned. Nancy thought they would have to cancel it, but everyone on Bigelow 13 urged them to go. "It's like Little Ireland there—Mary-Liz, Colleen Ryan, all the Irish nurses. Everyone said they thought it would be good for us." Tom's wounds were fully covered, the last bit closed on Nancy's birthday, July 16. She noticed that when she was taking off the Acticoat that morning, and she never put on another pad of it again. All Tom had to do was kick the Oxycontin. He thought it would be easy.

He figured he would do it cold turkey. "You know: Stop, get it over with, suck it up for two weeks. But it was a nightmare coming off that stuff." Ashley said it looked like he had a really bad case of the flu; Tom insisted it was worse than that. "It does a wonderful job, I can attest to that. But trying to come off it was just maddening. I couldn't sleep. I would just be lying there with my eyes open, just staring. After a while you want to go to sleep and you just can't. I literally watched the clock. . . . Eventually, gradually, my sleep returned. It was horrible." Tom shudders when he talks about it.

The ordeal lasted two weeks. At the end of it, Tom and Nancy boarded a plane for Ireland with a carry-on bag full of gauze, sterile water, Neosporin, and enough ibuprofen to get him through the pain.

Their wedding anniversary came during that horrible second week when Tom was in the hospital. "A really close friend of mine had called," Nancy remembered, "and when I told her, 'Tomorrow's our anniversary,' she started to cry and said, 'What a terrible anniversary.' I said, 'No, what a wonderful anniversary. He's alive. Six days ago I didn't know if he would be.'" While they were in Ireland, they renewed their wedding vows.

* * *

The outpatient clinic always had a makeshift look about it, tucked all the way back at the end of the corridor with the "waiting room" just some chairs pushed against the wall, and in front of those the desk where a receptionist sat, greeting people when they came in, checking off their names, calling them when it was their turn. The two examination rooms were just like the other rooms on the floor, except instead of hospital beds there was an examination table and a cabinet with bandages and tubes of ointment.

The people who come to clinic include former patients returning for monitoring and follow-up care, and people with less serious burns who do not need to be admitted to the hospital. They are seen by Mary-Liz, the residents, and the attending who is on that day. The staff works as a team, popping in and out of the examining rooms and conferring with each other in the hall. They examine wounds and change dressings, teach patients and family members how to do wound care, do debridement and other minor procedures, and keep an ever watchful eye and ear out for danger signs, be they physiological or psychological.

An outpatient with a large but superficial wound on one of his arms is back for the second time. As John Schulz starts to cut away the dressing to take a look at the man's arm, the woman who came in with him says he has been upset. Schulz asks if she is the man's wife. "No," she answers. "I'm his mother." He had a girlfriend, she explains, but she died in the same accident that burned him. He thinks it was his fault. She says they talked to "the nice lady" last time. Is she here today? Schulz puts down his scissors and pokes his head out into the hall. A moment later, Mary-Liz is in the room. She pulls up a stool and for the next ten minutes, the man has her undivided attention. She listens silently, her sapphire blue eyes fixed on him, as he chokes out his grief, his feelings of guilt, his pain. She excuses herself for a minute, and when she returns a page has gone out for Dr. Findley, who will be there by the time Mary-Liz finishes cleaning and rewrapping the arm.

Mary-Liz has an instinct for how people are feeling, and people open up to her. It does not have to be a patient. One day she was on her way to clinic when she spied the mother of two patients, siblings who had been in the same fire. One had been less badly burned and was already in rehab. The other was having a harder time of it and was still on Bigelow 13, though he was now out on the floor. Mary-Liz greeted her warmly, asking, "How's Spaulding?" The woman said, "All right," in a way you knew meant it really was not. Mary-Liz said, "Tell me," in a voice that invited the woman to speak her mind. "I guess I'm just spoiled from being here." Then she teared up and could not get out any more words. Mary-Liz turned on a dime. She put her arm around the woman and ushered her into a quiet corner, where the two sat down so Mary-Liz could listen while the woman poured out her heart. Whatever else Mary-Liz was supposed to be doing would have to wait. This was important, the most important thing going on right then, right there.

There are patients who need the clinic, emotionally as much as anything else, not just to keep a connection with the people who cared for them, but because it is a safe place, a more gentle transition back into their old lives, or the new lives they are trying to fashion.

In the spring of 2003, the clinic moved to shiny new quarters on the other side of the elevators from the main treatment area on Bigelow 13. It had clearly outgrown the cramped space on the floor, just as it outgrew the one afternoon a week it started with, then the full day, then two. In the new quarters, the clinic is open four days a week, with Mary-Liz taking Monday and Friday by herself. It means less time working with nurses as the clinical specialist, but more time doing what she likes best. "Taking care of patients gives me the most satisfaction."

For a while it looked like the clinic was going to move out of the building entirely, across Fruit Street to a new building that would eventually house all the MGH outpatient clinics. But staff and patients alike lobbied hard to keep it on Bigelow 13. Nancy Parent made her plea directly to the hospital administration. "I wrote a very long letter saying, 'Don't do it. Don't take it off the thirteenth floor. You don't understand what these people create.' Because the fact is that to this day, almost a year and a half later, we could walk in and everyone there is going to say, 'How are you? Let me see your arm.' Emotionally, it was like going home, it was a safe house, and it still is."

Tom was at the clinic every Thursday for the first six weeks after he was discharged, and Nancy came with him every time. It gave her plenty of opportunity to realize how much the place meant, not only to her but to the other patients and families in the hallway waiting room. "I think the hardest thing about going to clinic—or maybe it's the nicest thing about going to clinic, to give it a different perspective—is watching everyone interact with these folks with a 'Hello, you were a person before this happened, and you're a person now.' They're treated the same when you know for a fact that many of them don't experience that in the outside world. I don't care who you are, everyone's innate fear is burns: What if they were on fire, what if they were to burn." Almost reflexively, people avert their eyes. "It's kind of a natural reaction, but when you're there, it's just people."

Nancy remembers one woman in particular. She had been burned over an unbelievable percentage of her body and nearly died. She had been in the hospital for close to a year. "She came into the clinic and she was all out of breath, and they said to her, 'How come you're out of breath?' And she said, 'I ran up the thirteen flights.' And they said, 'Why'd you run up the thirteen flights?' And she looked at them, and she said, 'Because I can.'"

As much as the emotional and physical scars shadow burn survivors for the rest of their lives, few will say they are sorry to have been saved. Surprising? Maybe, but John Findley has observed that over and over again, even with the would-be suicides, most of whom seem to have burned their death wish out of their system. John Burke is convinced that even with significant disabilities, most burn survivors adapt and live fruitful lives. He and his research nurse, Peggy Haggerty, have been studying long-term quality of life in survivors of severe burns. "In the '70s, when we began to publish the fact that we could save people's lives that had 80 percent or more of the body surface burned, there was a very considerable, sometimes impolite, sometimes mean, comment around the country that what we were doing was saving people's lives to commit them to a life of misery and seclusion and that what we were doing was a very bad thing," he told me.

"Peggy had seen some of the people and realized that that was a lot of rubbish. So she decided that she would organize a long-term follow-up of the big burns, just to demonstrate what their quality of life was long term. We decided that it had to be at least ten years. So we started that in '74 and we finished in '89, and we're now in the process of writing that paper." Their conclusions: "They do remarkably well. Functionally, they are not all perfect, but you know, you adapt, and burn patients, when they survive, adapt enormously well. If you have a disability so your left hand doesn't work, you figure out how to do everything you want to do with your right hand. Even though if you measure their hand function, let's say, you can document that hand function is only 60 percent of what yours and mine is, their quality of life is the same as yours and mine because they have figured out how to solve that problem."

Clearly *futility* is not in John Burke's lexicon. "People are still screaming bloody murder. The number of people who aren't treated—what I used to call 'flowers at the bedside'; now it's 'comfort care'—in the burn business is enormous. And my personal view is that that's a waste of good folks."

* * *

After Tom and Nancy got back from Ireland, clinic visits were stretched out to once a month, and eventually every three, then every six months. Mary-Liz and the doctors wanted to keep tabs on how he was healing, to make sure nothing was infected, that he was regaining sensation and mobility in his hands, that he was not developing contractures. But there were other things to watch for too. To hear Tom tell it, Dr. Ryan had an eagle eye trained on every aspect of his recovery. One time she got on his case because his skin looked dry. Because there are no oil glands in scar tissue, keeping it moisturized is important. Another time she asked about his appetite. When Tom said it was good, she said, "Yeah, too good." Time to say good-bye to the free lunch, and the snacks and the seconds at mealtime. His wounds were closed and his metabolism back down to normal, so he no longer needed those extra calories. Toward the end of August, Tom was thinking it was time for him to go back to work, but when he came in for his next appointment and raised the issue with Dr. Ryan, Nancy remembers her saying, "Tell me again what you do for a living." It was September 18, 2001. "He told her he was a TV cameraman and she sat there for a minute and then she said, 'I can't let you go back.'" She probably knew as well as Tom did that the station would want to send him to Ground Zero and that he would not say no to the assignment. "She said, 'I can't let you be in the New York environment, not emotionally or physically.'"

As it happens, Dr. Sheridan was already in New York with a disaster medical assistance team (DMAT) that included Bob Droste and Sally Morton, as well as other MGH emergency personnel and burn specialists working at both Ground Zero and the Cornell Burn Unit at New York-Presbyterian Hospital. Bob and Sally spent two weeks lending a hand to the overwhelming task of caring for the burn victims of the World Trade Center attack.

No one was ignoring the emotional aspect of Tom's recovery. On an earlier visit, Nancy mentioned to Mary-Liz that Tom's even temper seemed to have gone missing. "There came a point where he was very angry all the time. It wasn't angry at me; he was just short fused. Tom's not a short-fuse guy. Tom's a 'water rolls off my back,' happy-go-lucky kind of guy." Mary-Liz had been on the lookout for cues that Tom was about to "hit the wall." She knew how important it is to anticipate the moment. "I wanted to be able to intervene before that point came. A patient has to be ready to discuss and accept post-traumatic stress disorder," she explained. "The timing is critical to John Findley's being able to help, so knowing when to call him is important. Part of my routine assessment with everyone are seemingly innocent questions about eating, sleeping, nightmares. I'm looking for the clues that tell me the time is right for the referral." Nancy's observation about Tom's uncharacteristic anger was the tip-off. "Although he sees all the inpatients, the bulk of Dr. Findley's work is done with patients after discharge. That's when it usually hits—big time."

Looking back on it, both Tom and Nancy know the entire first year was an emotional seesaw, between his adjustment and her conviction that it was all her fault. The true emotional impact of what had happened hit

Tom unaware. "I was feeling great for a long time, and then there was a woman at work whose father was in the Pentagon on the eleventh and it took a day to find him, and he had third-degree burns over 70 percent of his body, and he lived through it. It was when I was helping prepare her that all that posttraumatic stress just came right back. And that's when I fell apart."

Tom returned to his job just before Thanksgiving, easing his way in by working in the studio for a while before going back out into the field. He was no longer having nightmares every night, he was no longer in pain. It seemed like things were returning to normal.

Nancy knew better. "Tom was getting very anxious when we got to the one-year anniversary. I could see it happening, I could see it two months before, six weeks before. I said to him, 'You know what we're going to do. On May 26, we're going to have a huge party. And after May 26, 2002, we're not going to call ourselves burn victims anymore. We are now burn *survivors*.' And that's what we did. We had this huge party and invited everyone who had anything to do with us over the past year. And he was actually grilling.

"I really do think some of the perspective changed. We're not going to let this be an identifier going forward. It is something that happened, and we got beyond it."

* * *

Once burn survivors leave the safety of the hospital, they enter a world that is no more ready for them than they are for it. Unlike survivors of many other types of accidents and illnesses, they wear the story of their ordeal for all to see. The lucky ones, like Tom Parent, have the support of loving families or find strength in a religious or therapeutic community, as Dan O'Shea did. Many owe their emotional recovery to the Phoenix Society for Burn Survivors, a national organization dedicated to support, public education, and advocacy. Attendees at the society's annual World Burn Conference gather for workshops and lectures ranging from the latest in prosthetics and reconstructive techniques to issues of intimacy and sexuality, self-esteem, and coping with social and workplace situations. Unlike any other group of medical professionals, the American Burn Association includes burn survivors in its membership.

Attendance at the annual ABA meeting is always an eclectic mix of doctors, nurses, and others in the burn treatment community, firefighters and emergency medical technicians, and burn survivors, whose voices are heard and faces are seen by the people who saved their lives. Burn camps for children and retreats for adults or entire families provide a safe place as survivors prepare to reenter society. Local and Internet support groups continue to provide information, coping strategies, and above all a sense of community.

A bond forged by fire joins former burn victims. They share something that no one else understands and most people do not want to imagine. "It becomes almost like a little secret society when you find somebody that's been burned," Tom Parent told me. "Immediately you have a connection. It's almost like when you see two Vietnam vets talking, and they're in their own world. They know something you can't be a part of because you didn't experience it. If somebody asks me, I'll do my best to explain about burns and how it works, and why you should never be burned, but until you've been there, I don't care how I describe it, you will never capture the sensations and the feelings of what you go through when you get burned. I remember when 9/11 happened, and people were saying to me, 'I can't believe people were jumping,' and I said, 'Yeah, I can. If I had my choice between the two, guess where I'm gonna go.' And they'd look at me. But you're not going to understand it, nor do I ever wish you to understand it."

Tom and Nancy talked to me in the living room of their home on Cape Cod, on the other side of the wall from the kitchen where it all happened. The dog was dozing at Nancy's feet. It turns out you are a pretty bad shot when your hands are on fire. When Tom let go of the pot of boiling oil, it fell far short of both the doorway and the dog. The flames sputtered out as they met the concrete floor of the porch.

Tom squirmed a bit when Nancy talked about how heroic he was. "I'm just an arm," he said. "I have incredible respect and admiration for anybody who has more burn than I have. I can't even fathom what they went through, nor would I even try. I think of the old saying: The Lord will only give you as much burden as you can bear. So, okay, this is all I can bear, and I feel like a little wimpy guy compared to them."

* * *

When Dan O'Shea had been sober a year, he invited Mike Wilson to come to the meeting at which he would celebrate his anniversary. Mike wanted to, but something came up and he could not make it. He feels bad about that, but the truth is he rarely keeps in touch with patients after they have left Bigelow 13. Dan understands that, feels the same way himself. For a while, even after he no longer needed to come for clinic, he would stop by from time to time just to say hello, but then he knew it was time to let go. "I was very attached to Michael when I was there," Dan said. "But you know, I don't need and I don't think they need to hold onto a personal relationship. They all have a job to do, and they're very good at it, so probably the best thing for them is just to know that I am someone who was once there and now I'm doing well. It must be wonderful for them to see people fully recovered, but they all have lives of their own."

And that is absolutely true. Mike Wilson attains a kind of grace through his work. "It's a blessing to work here," he told me. He puts in three twelve-hour shifts a week, but he leaves it all behind when he goes home. While he is on Bigelow 13, there is no limit to what he will do for his patients and their families, but "There is a limit," he insisted. "When you're better, it's time for you to leave."

Notes

My principal sources throughout this book were the burn patients who allowed me to intrude upon their privacy, and the burn care professionals who allowed me to pick their brains and watch them work. Unless otherwise noted, all direct quotations, observations, thoughts, and feelings attributed to these people were drawn from in-person conversations—including formal sit-down interviews and informal exchanges in the course of a tour or a surgical or bedside procedure, casual discussion, and, in a few instances, out-and-out eavesdropping. All are referred to in these notes as interview. For information about the pathophysiology and treatment of burns, my principal source was the comprehensive and authoritative *Total Burn Care*, ed. D Herndon, 2d ed. (Philadelphia: Saunders, 2002), which is referenced throughout these notes as *TBC*. Additional sources that turn up repeatedly are the American Burn Association, "Practice Guidelines for Burn Care," *J Burn Care Rehabil*, April 2001, here referred to as PGBC; and, as a source of statistics on burn incidence and outcomes, National Burn Repository, *2002 Report* (Chicago: American Burn Association, 2002), here referred to as NBR.

Introduction

xi   *The earliest known writings:* Details and translation from the Ebers Papyrus: J Scarborough, "Medications for Burns in Classical Antiquity," *Clin Plas Surg* 10 (1983): 603–610.

xv   *"Burns have a personality all their own":* Mike Wilson, interview, October 7, 2002.

xvi   *the author was Jacob Bigelow:* Jacob Bigelow (1787–1879) was a major figure in the history of American medicine. In addition to his twenty-year association with Massachusetts General Hospital, he was a professor at Harvard Medical School and served as a president of the Massachusetts Medical Society. An avid botanist, he was a member of the committee that established the *American Pharmacopoeia*, the official encyclopedia of drugs and medicinals, and is recognized for developing the system of nomenclature most widely used in the *materia medica* (pharmacology). He also happened to be the force behind the design and building of Mount Auburn Cemetery, America's first garden cemetery and now a national historic landmark. Bigelow's son, John Jacob, followed in his father's footsteps and had a distinguished career at both Harvard Medical School and Mass General. He was present when Dr. John Collins Warren performed the first surgical procedure using general anesthesia, an event commemorated by a mural in what came to be called the Ether Dome, the hospital's first operating theater. The site has been restored as a museum open to the public in the historic Bulfinch building, the hospital's original structure and the centerpiece of the present-day complex. The younger Bigelow's account, "Insensibility During Surgical Operations Produced by Inhalation," was published in the *Boston Medical and Surgical Journal*, November 18, 1846. Both father and son are buried in Mount Auburn Cemetery, along with Oliver Wendell Holmes, Henry Wadsworth Longfellow, and a host of Cabots and Lowells, among other eminent Bostonians.

xvi   *Observing that "the distressing":* J Bigelow, "Observations and Experiments on the Treatment of Injuries Occasioned by Fire and Heated Substances," *New England Journal of Medicine and Surgery and the Collateral Branches of Science* I (1812): 52–64.

Prologue

1   *Sometime in the early hours:* The events described are all true, the thoughts are those recollected and communicated by the participants in a series of interviews between August and October 2002. The sequence of events was pieced together with the help of the people known in this book as Dan, Peggy, and Jack O'Shea, and the unfailing patience of John Schulz, MD, who had access to Dan's hospital records and Dan's kind permission to share information contained in them.

Chapter 1: Saturday Night at the Cocoanut Grove

3   *was packed to the rafters:* The Cocoanut Grove fire is still remembered by Bostonians, including a few who were there and many others with relatives or friends who survived or died in the fire. The story of the fire and its af-

termath has been told in two books: Paul Benzaquin, *Fire in Boston's Co-coanut Grove: Holocaust!* (Boston: Branden, 1967); and Edward Keyes, *The Cocoanut Grove* (New York: Atheneum; 1984). Both are out of print but widely available in libraries. Innumerable newspaper articles were published at the time and in 1992 to coincide with the fiftieth anniversary of the fire. What is less well-known to the general public is how victims were cared for once they reached Boston City and Massachusetts General hospitals. This treatment was thoroughly documented by scores of papers written in the years following the fire and published in various medical journals. The most comprehensive record of the Mass General experience can be found in "Symposium on the Management of the Cocoanut Grove Burns at the Massachusetts General Hospital," *Ann Surg* 117 (1943):801–965; hereafter Symposium. The entire issue of the journal consisted of sixteen articles covering every discipline, from social services to surgery, and every stage, from admission to autopsy. Papers coming out of Boston City Hospital were published separately in numerous journals, on topics ranging from the surface treatment of burns (GHA Clowes, CL Lund, SM Levenson, "The Surface Treatment of Burns," *Ann Surg* 118 [1943]:761–779) and infection control, to shock, renal failure, wound healing, metabolic and nutritional problems, and neuropsychiatric issues, the last including Alexandra Adler's landmark study of posttraumatic stress disorder ("Neuropsychiatric Complications in Victims of Boston's Co-coanut Grove Disaster," *JAMA* 123 [1943]: 1098–1101). A monograph that is considered the seminal study of inhalation injury (M Finland, CS Davidson, SM Levenson, "Clinical and Therapeutic Aspects of the Conflagration Injuries to the Respiratory Tract Sustained by Victims of the Co-coanut Grove Disaster," *Medicine* 25 [1946]: 215–283; hereafter Finland, Davidson, Levenson) was one of the principal sources for this chapter.

Additional information was drawn from eyewitness recollections published in the *Harvard Medical Alumni Bulletin,* Winter 1991–1992: 10–19; FD Moore, "Death After the Game: The Cocoanut Grove Fire," in *A Miracle and a Privilege* (Washington, D.C.: Joseph Henry Press, 1995), chap. 9; SM Levenson, "Thermal Burns: A Personal Odyssey," *Surg Clin N Am* 58 (1978): 1005–1017; an amateur videotape of a 1985 symposium on the fire held at Boston City Hospital and attended by a group of BCH physicians who had been there at the time of the fire and, from MGH, Oliver Cope (hereafter video); and Stanley Levenson's personal reminiscences recorded during telephone interviews on October 14, 2001, and July 2, 2002.

5   *death toll stood at 492:* The most often cited contemporary account (RS Moulton, "The Cocoanut Grove Night Club Fire, Boston, November 28, 1942," National Fire Protection Association, Boston, January 11, 1943: 1–19) puts 200 "victims" at the main entrance and 100 more at the exit from the cocktail lounge. This number seems not to include those lost in the crush in the basement. With more than 300 dead at the scene and 300 DOA at the hospitals, it is hard to make these figures add up to only 491 or 492. The discrepancy may be due in part to vagueness in Moulton's description—the piles of bodies may have been a mix of the living, the dead, and the dying, and some of each were undoubtedly sent to hospitals, where they were more definitively sorted out.

6   *negligence and deliberate violations:* Despite numerous violations, the Cocoanut Grove had recently passed a fire department safety inspection. An investigation after the fire revealed a web of bribery and corruption that reached as far as City Hall. Ten people were indicted on various charges, but the club's owner was the only one who went to trial. He was convicted of manslaughter and sent to prison for life, a sentence cut short by his death from cancer in 1947.

6   *According to one account:* Moulton, "Cocoanut Grove Night Club Fire."

7   *Harvard, Tufts, and Boston University:* Boston City Hospital no longer exists, at least not by that name. In the 1970s, the Harvard and Tufts residency programs were discontinued and the hospital itself was absorbed into what is now called Boston Medical Center, which hosts the Boston University residency program exclusively. Mass General remains the principal location for Harvard's residency program. Tufts residents train at New England Medical Center.

7   *"had made us catastrophe-minded":* Cope, foreword to Symposium: 801–802.

8   *three men with burns:* Based on an eyewitness account by Dr. Crawford Adams, a Boston University medical intern, quoted in Finland, Davidson, Levenson.

8   *"The examining rooms":* Finland, Davidson, Levenson.

8   *Stanley Levenson was a young burn fellow:* He went on to a long and distinguished career as a surgeon, researcher, and teacher. Now University Professor Emeritus at Albert Einstein College of Medicine, he was for many years director of the surgical ICU/burn unit at Bronx Municipal Hospital Center. Nearing ninety, he is still active professionally. The two men with whom he worked so closely, Maxwell Finland and Charles Davidson, died in 1987 and 2000 respectively.

8   *"in vivid detail":* Unless otherwise noted, all quotations from Stanley Levenson in this chapter are taken from telephone interviews on October 14, 2001, and July 2, 2002.

10 *the overwhelming majority:* Finland, Davidson, Levenson; JC Aub, H Pittman, AM Brues, "The Pulmonary Complications: A Clinical Description," in Symposium: 834–840.

10 *classic study of inhalation injury:* Finland, Davidson, Levenson.

10 *"We spent hours":* This recollection and all others attributed to Charles Davidson in this chapter have been transcribed from video.

11 *chart and accompanying schematic diagram:* Table 6 and figure 1: "Schematic Representation of Location of the Patients at the Time They First Became Aware of the Fire and the General Direction of Their Attempted Escape," Finland, Davidson, Levenson.

12 *"all parts of the building":* Finland, Davidson, Levenson.

12 *"severest respiratory tract damage":* Finland, Davidson, Levenson.

13 *the Rialto Theater fire:* Seven people were killed, but the eighty who were injured provided Underhill with the opportunity to study burn shock firsthand and publish his findings in "The Significance of Anhydremia in Extensive Surface Burns," *JAMA* 95 (1930): 852–857.

14 *had in common:* Finland, Davidson, Levenson.

14 *In the end . . . in the first two or three days:* Finland, Davidson, Levenson.

14 *some degree of respiratory injury:* Finland, Davidson, Levenson; NW Faxon, "The Problems of the Hospital Administration," in Symposium: 803–808; Aub, Pittman, Brues, "Pulmonary Complications," in Symposium.

15 *1,000 cc of fluid:* Faxon, "Hospital Administration," in Symposium; O Cope, FW Rhinelander, "The Problem of Burn Shock Complicated By Pulmonary Damage," in Symposium: 915–928.

15 *afraid that giving more:* Cope, Rhinelander, "Burn Shock," in Symposium; Finland, Davidson, Levenson.

15 *"decidedly different at the two institutions":* Finland, Davidson, Levenson.

17 *Cope dated . . . "Unforgettable":* This is a story Cope was fond of telling. Most of this version came from a 1983 interview excerpted in "The End of the Tannic Acid Era," *Harvard Medical Alumni Bulletin*, 17–19. He can also be seen telling it on video. By that time, Cope was a rather grand old man. Tall, elegant, with chiseled features and snow white, bushy eyebrows that formed a shelf above his eyes, he spoke with the slightly British accent affected by highly educated Bostonians, striking his fist against the lectern to emphasize his outrage.

18 *"what healed most rapidly":* Cope, quoted in "End of the Tannic Acid Era."

18 *They completed their study:* The work was done before the Cocoanut Grove fire but not published until a few months after it. B Cannon, O Cope, "Rate of Epithelial Regeneration: A Clinical Method of Measurement, and

the Effect of Various Agents Recommended in the Treatment of Burns,"
Ann Surg 117 (1943): 85–92.

18 *"We had proved convincingly":* B Cannon, "What We Learned," *Harvard
Medical Alumni Bulletin:* 18.

18 *toxic to the liver:* RD McClure, CR Lam, H Romence, "Tannic Acid and
the Treatment of Burns: An Obsequy," *Ann Surg* 120 (1944): 387–398.

18 *In his report:* Cope, "Treatment of the Surface Burns," in Symposium.

18 *"The treatment used . . . when not to interfere":* Cope, "Treatment of the
Surface Burns," in Symposium.

19 *the U.S. Army protocol:* Clowes, Lund, and Levenson's study compared a
variety of surface treatments—those used in the Cocoanut Grove patients
as well as Cope's gauze and ointment method, which BCH started using
after the February 1943 publication of the Symposium. GHA Clowes, CL
Lund, SM Levenson, "The Surface Treatment of Burns," *Ann Surg* 118
(1943):761–779; "Guides to Therapy for Medical Officers," War Department, TM8–210, March 20, 1942.

19 *"supplies of silver nitrate":* Clowes, Lund, Levenson, "Surface Treatment."

19 *A study out of BCH:* Clowes, Lund, Levenson, "Surface Treatment."

20 *formula for calculating:* O Cope, FD Moore, "The Redistribution of Body
Water and the Fluid Therapy of the Burned Patient," *Ann Surg* 126
(1947): 1010–1045. Cope and Moore drew on the work of Alfred Blalock,
who demonstrated and quantified the fluid shift in burns, using a dog animal model. Blalock's pioneering studies, first at Vanderbilt University and
later at Johns Hopkins, revolutionized the understanding and treatment of
shock. He in turn was influenced by Walter B. Cannon, the renowned
Harvard physiologist who studied traumatic shock during World War I
and was the father of Bradford Cannon, the plastic surgeon with whom
Cope proved the shortcomings of tannic acid and triple dyes. Blalock is
best remembered, however, as the man who, along with Helen Taussig and
his brilliant lab technician Vivien Thomas, devised the "blue baby" surgery
that saved the lives of tens of thousands of children born with heart defects
and ushered in the modern age of cardiac surgery.

20 *should be removed surgically:* O Cope, JL Langohr, FD Moore, RC Webster, "Expeditious Care of Full-Thickness Burn Wounds by Surgical Excision and Grafting," *Ann Surg* 125 (1947): 1–22. Oliver Cope was a
towering figure. As a young surgeon and researcher, he did important
work on the metabolic effects of diseases of the thyroid and parathyroid
glands. At the end of his career and before his death in 1994, he champi-
oned breast-sparing surgery as a safe and effective alternative to radical
mastectomy in the treatment of breast cancer. But it was in midcareer he

made his mark in the world of burns. Like many innovative thinkers, he saw connections that others missed.

Chapter 2: A Riot in the Body

22 *catastrophic nightclub fire:* As of this writing, the story of the fire is still unfolding. Details were gleaned from television, radio, and newspaper accounts, press briefings and updates from Rhode Island and Massachusetts General hospitals, and interviews with staff at Mass General.

23 *passed a fire inspection:* Under existing fire codes, the club did not need a sprinkler system for reasons given variously as its small size and a grandfather clause exempting it from this safety feature. It was widely reported that fire officials had found other violations in earlier inspections and ordered that they be corrected. On December 31, 2002, the club was found to be in compliance. At no time was note taken of the foam, which was never intended to be used as soundproofing, was not rated as flame retardant, and had been attached to the wall with an aerosol adhesive that was itself a powerful accelerant once the fire started.

23 *by a local television crew:* In an even more bizarre twist, the crew included one of the owners of the club, who was also a reporter for the television station and was doing a story on nightclub safety as a follow-up on the Chicago tragedy earlier in the week.

24 *more than enough blame:* In December 2003, criminal charges were brought against only three people in connection with the fire. The two brothers who owned the nightclub and the tour manager of the band were each indicted on 200 counts of involuntary manslaughter—one charge of criminal negligence and one of misdemeanor manslaughter for each of the 100 dead. Many survivors, victim families, and community members were outraged that more people were not deemed culpable. In time, civil suits may widen the net.

24 *Injuries . . . numbered close to 200:* It was initially reported that 187 people had been taken to hospitals, though many were treated and released. An untold number may have been injured slightly enough that they did not seek medical attention.

26 *Medical science took a long time:* An excellent gloss as well as a fascinating read can be found in WP Cockshott, "The History of the Treatment of Burns," *Surg Gynecol Obst* 102 (1956): 116–124.

26 *"if the burn is superficial":* Quoted in E Vander Elst, "Historical Aspects of the Treatment of Burns," in *Physiopathology and Treatment of Burns,* ed. J Lorthior (Brussels: Presses Academiques Europeenes, 1964), 1–23.

26 *ample opportunity to study:* Paré chronicled his battlefield experience in *Journeys in Diverse Places.* This account includes the oft quoted remark, "I dressed him, and God healed him." A translation by Stephen Paget, originally published in Paget's own book, *Ambroise Paré and His Times* (New York: Putnam's, n.d.), and anthologized in *The Harvard Classics: Scientific Papers (Physiology, Medicine, Surgery, Geology),* ed. CW Eliot (New York: Collier, 1909–1914), vol. 38, can be found online: Project Gutenberg www.ibiblio.org/gutenberg/etext04/hvrcl10.txt; and www.bartleby.com/38/2/1.html.

26 *"Black skin . . . violence of the fire"* and *"It is easy to treat burns . . . severe symptoms follow":* JJR Kirkpatrick, B Curtis, M Fitzgerald, IL Naylor, "A Modern Translation and Interpretation of the Treatise on Burns of Fabricius Hildanus (1560–1634)," *Br J Plas Surg* 48 (1995): 460–470.

27 *"retracted sinews":* IL Naylor, B Curtis, JJR Kirkpatrick, "Treatment of Burn Scars and Contractures in the Early Seventeenth Century: Wilhelm Fabry's Approach," *Medical History* 40 (1996): 472–486.

27 *Dupuytren maintained there were six degrees:* HN Harkins, *The Treatment of Burns* (Baltimore: Thomas, 1943), introduction.

27 *article on burns and scalds:* This article can be found in the online electronic version of the *Encyclopedia Britannica,* 11th ed. (1911): http://100.1911encyclopedia.org/B/BU/BURNS_AND_SCALDS.htm.

30 *devised a way to map burns:* CC Lund, NC Browder, "The Estimation of Areas of Burns," *Surg Gynecol Obstet* 79 (1944): 352–358.

31 *"burns are not a simple problem":* J Long, "Post Mortem Appearances Found After Burns," *Lond Med Gaz* 25 (1840): 743–750, quoted in HK Hawkins and HA Linares, "The Burn Problem: A Pathologist's Perspective," in *TBC,* chap. 40.

33 *the only particles too large:* "The increased permeability permits molecules of up to 350,000 molecular weight to escape from the microvasculature, a size which allows essentially all elements of the vascular space except red blood cells to escape from it." "Burn Shock Resuscitation: Initial Management and Overview," in PGBC, chap. 7.

33 *the swelling concentrates . . . the capillary leakage spreads:* GC Kramer, T Lund, DN Herndon, "Pathophysiology of Burn Shock," in *TBC,* chap. 7.

34 *smallest burns:* Different authorities give different burn sizes in which edema spreads beyond the local burn, ranging from 10 percent or larger to 25 percent or larger.

37 *He is quoted:* Harkins, *Treatment of Burns,* introduction.

38 *Oliver Cope and Francis Moore's determination:* "The Redistribution of Body Water and the Fluid Therapy of the Burned Patient," *Ann Surg* 126 (1947): 1010–1045.

38   *Different fluids and combinations:* The formula developed by Everett Idris Evans in 1952 used 1 milliliter each of normal saline and colloid per kilogram of body weight times percentage TBSA; in 1953, the Brooke formula replaced the normal saline with lactated Ringer's and altered the proportions: 1.5 ml lactated Ringer's and 0.5 ml colloid per kilogram times percentage TBSA; the Brooke formula was later modified to eliminate colloid entirely, so the full 2 ml/kg/% TBSA consisted of lactated Ringer's.

In 1951, M. J. Kyle and A. B. Wallace established that fluid replacement requirements for children differ markedly from those of adults and thus resuscitation formulas should differ as well. Greater precision is needed to prevent shock in children. In addition, children require relatively more fluid than adults for burns of the same size, and need to be resuscitated for small burns (10%–20% TBSA) whereas an adult would not. The formulas for children use lactated Ringer's, with the addition of sodium bicarbonate and/or albumin. S Thomas, RE Barrow, DN Herndon, "History of the Treatment of Burns," in *TBC*, introduction; "Burn Shock Resuscitation: Initial Management and Overview," in PGBC, chap. 7.

39   *probably do more harm than good:* WW Monafo, "Initial Management of Burns," *N Engl J Med* 335 (1986): 1581–1586.

39   *combine lactated Ringer's:* In the 1970s, William Monafo advanced the theory that adding more sodium would make it possible to use less fluid and thus reduce edema formation. The formula he proposed used 1–2 ml/kg/%TBSA of a hypertonic saline solution (250mEq Na/L, milliequivalents per liter of water) over twenty-four hours. The benefit of using less fluid, however, may not balance against the risk of excess sodium levels in the blood (hypernatremia). Other hypertonic solutions use less sodium. The Warden formula uses a combination of lactated Ringer's and sodium bicarbonate for the first eight hours, then lactated Ringer's only for the second eight hours in an amount that maintains urine output (the output goal is usually 0.5 milliliters per kilogram per hour), and then Ringer's with the addition of albumin for the third eight-hour period, again with urine output guiding the volume.

The principal colloid formulas use lactated Ringer's (2 liters over 24 hours) with fresh frozen plasma with or without the addition of dextran in varying amounts, with monitoring of urine output. Thomas, Barrow, Herndon, "History of the Treatment of Burns"; "Burn Shock Resuscitation."

39   *"It is clear":* "Burn Shock Resuscitation."

40   *there is not much he doesn't know:* Bob Droste interviews, October 7, 2002; May 5, 2003.

43 *accelerated metabolic rate and increased energy demand:* M Spies, MJ Muller, DN Herndon, "Modulation of the Hypermetabolic Response After Burn," in *TBC*, chap. 27.

45 *both Boston City and Mass General:* Nutrition was a major interest of Charles Lund and a focus of research for both Charles Davidson and Stanley Levenson at the Thorndike lab. Dr. Lund devised a liquid feed mixed up in a blender at BCH to make it easier for both nurses and burn patients, as well as to more closely monitor what and how much the patients were taking in. Papers on nutrition and metabolism that came out of the BCH experience included, among others, FHL Taylor, SM Levenson, CS Davidson, MA Adams, "Abnormal Nitrogen Metabolism in Patients with Thermal Burns," *N Engl J Med* 229 (1943): 855–859; FHL Taylor, SM Levenson, CS Davidson, NC Browder, "Problems of Protein Nutrition in Burned Patients," *Ann Surg* 118 (1943): 215–224; FHL Taylor, SM Levenson, MA Adams, "Abnormal Carbohydrate Metabolism in Human Thermal Burns," *N Engl J Med* 231 (1944): 437–445. Oliver Cope and Francis Moore also published important work in this area, beginning with O Cope, IT Nathanson, GM Rourke, H Wilson, "Metabolic Observations," in Symposium, 937–958, and continuing into the 1950s with O Cope et al., "Metabolic Rate and Thyroid Function Following Acute Thermal Trauma in Man," *Ann Surg* 137 (1953): 165–174; FD Moore, *Burns in Metabolic Care of the Surgical Patient* (Philadelphia: Saunders, 1959).

45 *"the hallmark of a burn injury":* JF Burke, "Fluid Therapy to Reduce Morbidity," in NI General Medical Sciences with ABA, Consensus Development Conference on Supportive Therapy in Burn Care, *J Trauma* 19, suppl (1979): 865–866.

46 *139 specialized burn centers:* The number of specialized burn centers verified by the American Burn Association and the American College of Surgeons is in flux, but the trend is downward. According to the 2002 presidential address given by outgoing ABA president Jeffrey R. Saffle, MD, there were 139 in 2000, down from 185 in 1981. Burn centers have to undergo reverification every three years.

Chapter 3: The General

47 *It was 4:52 Saturday morning:* The sequence of events described in this chapter, beginning with Dan O'Shea's arrival at the MGH Emergency Department, is re-created from John Schulz's recollections of that day, aided by his review of Dan's hospital record, and communicated in interviews on October 10 and November 15, 2002, and subsequent e-mails. General in-

formation about operations and procedures were provided during a tour of the ED on October 8, 2002, given by Kelley Burke, RN, who has worked as a trauma nurse in the ED since 1989 and before that worked in the adult burn unit on Bigelow 13.

47   *Whenever there is a fire:* Details of the Turret operations and CMED dispatch procedures provided in a telephone interview on December 16, 2002, by a spokesperson for Boston Emergency Medical Services, which is under contract with the city of Boston to manage the CMED system.

48   *emergency rooms . . . filled to overflowing:* Boston is not unique in this respect. In too many large urban areas, emergency and other hospital-based facilities are stretched to the breaking point. Hospital consolidations and closings, compounded by the staffing shortages that are endemic across the health care professions, mean fewer personnel and beds for the growing number in need of care, many of whom live in medically underserved areas and are themselves under- or uninsured. Guaranteed emergency medical treatment may be as close as the United States will ever get to universal health care, but it is also the reason why the nation's overcrowded hospital emergency rooms, with their overtaxed personnel, have become the gateway to medical treatment for all too many Americans. Changes in the Emergency Medical Treatment and Labor Act (EMTALA) of 1985, which was originally intended to prevent patient "dumping" by medical facilities unwilling to give appropriate emergency treatment to uninsured patients, are likely to exacerbate the situation, despite the stated intention of "clarification" and "removing barriers to the efficient operation of hospital emergency departments" (press release, Center for Medicare and Medicaid Services, August 29, 2003). The net effect of the rule, effective November 10, 2003, is to reduce the number of on-call specialists available to those seeking emergency care. Anyone who wishes to gaze upon the apocalyptic future of medical care in the United States need look no farther than the waiting area outside the emergency room of any large metropolitan hospital.

49   *American College of Surgeons Committee on Trauma:* These standards include a lengthy description of required personnel, equipment, case volume, facilities, governance, and review, as set forth in the ASCOT publication, "Resources for Optimal Care of the Injured Patient" (Committee on Trauma, American College of Surgeons, 1999), which was originally published under the title "Optimal Hospital Resources for Care of Injured Patients," *Bull Am Coll Surg* 64 (1979): 43–48. It has undergone periodic revisions and name changes in the intervening years. Burn center verification is jointly administered by ASCOT and the American Burn Association to ensure compliance with guidelines to optimize burn care.

50 *advanced trauma life support:* ATLS is one of a series of emergency medical protocols for managing life-threatening conditions. Training is offered to doctors, nurses, emergency rescue and ambulance personnel, firefighters, and other first responders by the appropriate medical specialty. For example, ATLS is given as a continuing medical education course by the American College of Surgeons, advanced cardiac life support (ACLS) is sponsored by the American Heart Association, advanced pediatric life support (APLS) is offered by the American Academy of Pediatrics, and advanced burn life support (ABLS) by the American Burn Association. Other courses are available in basic life support (BLS), required for all EMTs; advanced life support (ALS), required for all paramedics; and advanced hazmat life support (AHLS), which covers the medical management of patients exposed to toxic and other hazardous materials.

55 *Peggy O'Shea was too distraught:* The events and emotions of that first day are drawn from the recollections of Peggy and Jack O'Shea, interview, September 18, 2002.

Chapter 4: The Very Young, the Very Old, the Drunk, and the Stupid

62 *half of all burns . . . motor vehicle accidents:* Figures 15 and 16, "Place of Occurrence," in NBR; Facts on Fire, U.S. Fire Administration, Federal Emergency Management Agency, www.usfa.fema.gov/dhtml/public/facts.cfm; Fire Death and Injuries, National Center for Injury Prevention and Control, Centers for Disease Control and Prevention, www.cdc.gov/ncipc/factsheets/fire.htm.

62 *Most of it falls on Carla Cucinatti:* And throughout chapter, interview, June 20, 2002.

63 *Most burns are accidental:* Figures 17 and 18, "Circumstances of Injury," in NBR.

63 *"I wish we could go on national TV":* Interview, October 7, 2002.

63 *According to Dr. Findley:* And throughout chapter, interview, July 16, 2002.

65 *"You have to hack off":* Bob Droste, interview, October 7, 2002.

65 *The peak ages for getting burned:* Figure 6, "Age Group," in NBR.

65 *patients between the ages of five and thirty . . . disabled by their injuries:* BA Pruitt, CW Goodwin, AD Mason, "Epidemiological, Demographic, and Outcome Characteristics of Burn Injury," in *TBC*, chap 2; table 5: "Lived/Died by Age Group"; table 10: "Mortality Rate by Age Group and Burn Size"; table 21: "Days per % TBSA and Charges per Day by Age Group and Survival," in NBR.

68   *The highest incidence of nonaccidental burns:* Pruitt, Goodwin, Mason, "Epidemiological, Demographic, and Outcome Characteristics. "

68   *It does not take long: Scalds: A Burning Issu*e (Chicago: American Burn Association, 2000); "Hot Water Burns," National AG Safety Database, www.cdc.gov/nasd/docs/d000701-d000800/d000702/d000702.html.

69   *Child Abuse Prevention and Treatment Act (CAPTA):* U.S. Code Title 42, chap. 67.

70   *It costs a lot:* Figure 22: "Hospital Charges by Etiology"; table 18: "Hospital Charges," table 19: "Hospital Charges: Lived/Died by Burn Size Group"; table 21: "Days per % TBSA and Charges per Day by Age Group and Survival," in NBR; Cathy DeVitto, financial services counselor at MGH, telephone interview, January 24, 2003.

70   *health insurers and managed care organizations:* The exception is the hospitals for burned children endowed by the Ancient Arabic Order of the Nobles of the Mystic Shrine, a fraternal organization of mostly white men with absolutely no connection to Arabs or Islam but with a commitment to philanthropy, particularly for sick children. At Shriners hospitals, no one, rich or poor, is ever billed for care. There are four Shriners burns hospitals in the United States—in Galveston, Cincinnati, Sacramento, and Boston, a block away from MGH. From its founding in 1968, the thirty-bed Boston facility has been affiliated with the General, sharing its medical director as well as some of its surgical staff. John Schulz and his partners, Colleen Ryan and Rob Sheridan, care for patients in both places. Adult burn victims who are brought to the ED at the General are treated on Bigelow 13; those under eighteen are sent to the Shrine.

71   *Every effort is made:* Katherine Flaherty, director of Medicaid and uncompensated care programs for Partners Health Care, telephone interview, January 27, 2003.

71   *topping $20 million annually:* MGH *2002 Annual Report,* Massachusetts General Hospital.

71   *No one is getting rich:* In his presidential address to the American Burn Association, Jeffrey R. Saffle, MD, put it succinctly: "Our patients are unpopular because they're a lot of trouble, they cost a lot of money, and they have visible, disturbing injuries. Our specialty is unpopular because it's labor-intensive, far from glamorous, less remunerative than many others." Saffle, "The 2002 Presidential Address: N.P.D.G.B. and Other Surgical Sayings," *J Burn Care Rehabil* 23 (2003): 375–384.

71   *"There are people who think what we do is a sin":* John Schulz, interview, November 15, 2002.

72   *"When there are bad times":* Lynn Bellavia, interview, March 6, 2002.

72   *the case of Tom Parent:* The sequence of events, as well as the thoughts at-
      tributed to members of the Parent family, are drawn from an interview on
      October 27, 2002, with Tom, Nancy, and Ashley Parent in their Cape Cod
      home, the scene of the fire.

76   *sent to a specialized burn unit:* According to criteria developed by the Amer-
      ican Burn Association and the American College of Surgeons, burn victims
      should be transferred if they have suffered partial-thickness (second-
      degree) burns that cover more than 10 percent of the total body surface or
      full-thickness (third-degree) burns of any size; burns that involve the face,
      hands, feet, genitals, perineum, or major joints; electrical burns, including
      lightning injury; chemical burns; inhalation injury; a burn plus one or
      more preexisting medical conditions that could complicate management,
      prolong recovery, or affect mortality; burns plus trauma (such as broken
      bones) in which the burn injury poses the greater immediate risk. Burned
      children in any hospital that lacks qualified personnel or equipment to care
      for children should be transferred to a burn unit, as should patients who
      require special social or emotional intervention, or long-term rehabilita-
      tion (adapted from *Resources for Optimal Care,* chap. 14).

Chapter 5: Bigelow 13

79   *Sumner Redstone Burn Center:* The center is named for the media mogul
      whose life was saved by Dr. John Burke and the burn team at Mass General
      after he sustained massive burns in a fire at Boston's Copley Plaza Hotel in
      1979. Redstone was originally taken to Boston City Hospital but trans-
      ferred to MGH, where he remained for many months. He sued the hotel
      for negligence and handed over the proceeds to the burn center. He also
      helped finance Burke's research and development of the groundbreaking
      skin substitute, Integra. Redstone tells the story of the fire and his treat-
      ment and long recovery at MGH in the prologue to his book, *A Passion to
      Win,* with Peter Knobler (New York: Simon & Schuster, 2001).

79   *the outpatient clinic:* In the spring of 2003, the outpatient clinic was moved
      to a separate area adjoining the treatment floor. The old examination
      rooms at the end of the corridor were turned into offices for the nurse
      manager, clinical nurse specialist, and research nurse. When Dan O'Shea
      and Tom Parent were regular visitors to the clinic, it was down the hall
      from the beds they occupied as inpatients.

80   *more patients than available beds:* When things get really tight, burn pa-
      tients are housed on other floors. After the Rhode Island fire, fourteen pa-
      tients in critical condition were sent to MGH. Bigelow 13 took four,

sending the remainder to the SICU and MICU at the General and, in this exceptional case, to Shriners. On the other hand, when burns has spare beds and other services have ICU needs, the unit will find itself with "boarder" patients whose problems have nothing to do with burns. The burn nurses take care of them anyway. That is the way it works: The nurses come with the beds.

80   "*I was the least worst*": Tom and Nancy Parent's accounts of the events and emotions during Tom's hospital stay; interview, October 27, 2002.

80   "*After we got married*": Frank Ireland, interview, June 21, 2002.

82   *sleep-deprived physicians-in-training:* After decades of being treated like galley slaves in a system defended by the medical establishment with the fervor of graybeards extolling the character-building value of long walks to school in hip-deep snow, residents have finally been given a break. Sort of. The Accreditation Council for Graduate Medical Education (ACGME), which gives its blessing to 7,800 graduate medical programs in 118 specialties and subspecialties, has declared traditional residents' working conditions hazardous to everyone's health. Specifically it points to "patient safety and resident well-being" as major concerns. After years of insisting only that resident work hours not be "excessive," the ACGME was moved to act amidst a groundswell of grumbling among the rank and file, finger-pointing by the respected Institute of Medicine and others, and several high-profile lawsuits brought against hospitals when major medical errors resulted, perhaps, from the misjudgment of exhausted residents. As of July 2003, residents' hours in all accredited programs are limited to eighty a week with one day off every seven, and with no stretch of work to exceed twenty-four hours, to be followed by a mandatory ten hours off. Though down from the time-honored thirty-six-hour shifts and hundred-hour plus weeks, this is still no one's idea of a cushy job. As a consequence—intended or not—there are fewer resident hours to go around and the burn service, as well as other subspecialties, has less resident support than it had in the past. John Schulz explained in a written communication on October 30, 2003: "Burn experience is not a requirement for training in general surgery; subspecialties were cut in order to protect training time in general surgery." To take up the slack, each attending works an average of eight extra *scheduled* hours a week; unscheduled hours may stretch far beyond that. A full-time nurse practitioner, who can do much (though not all) of what the residents can do, has been added to share the burden.

82   *As Mike Wilson . . . put it:* Interview, June 21, 2002.

82   *John Schulz vigorously disputes:* Written communication, October 30, 2003.

83   *"If a nurse decides":* Mary-Liz Bilodeau, interview, March 6, 2002.

83   *sees it in a more positive light:* Unless otherwise noted, throughout chapter, John Schulz, interview, June 20, 2002.

83   *"One of the most notable characteristics":* John Schulz, written communication, October 22, 2003.

84   *Jennifer Verbesey:* Telephone interview, March 2, 2003. Dr. Verbesey was one of a special category of burn residents. Her residency is at the Lahey Clinic Medical Center, in nearby Burlington, Massachusetts. Because Lahey does not have a separate burn service, MGH takes its residents for the burn rotation during their second year.

86   *"In nonburn patients":* Jennifer Verbesey, bedside interview, June 21, 2002.

88   *"Let me start the conversation":* Nancy Parent recalled this conversation and the resident's words in interview, October 27, 2002.

90   *"We used to take rubber tubing":* And throughout chapter, Bob Droste, interview, May 5, 2003.

90   *"Flies were always":* John Burke, interview, April 9, 2002.

91   *his "abiding interest":* John Burke, interview, June 19, 2002.

92   *a stunningly simple design:* JF Burke, "Bacteria-free Nursing Unit: A New Approach to Isolation Procedures," *Hospitals* 43 (1969): 86–91.

Chapter 6: In the Hyperbaric Chamber

96   *hundreds of such facilities:* The Undersea and Hyperbaric Medicine Society, the accreditation authority for hyperbaric treatment centers, lists a total of 500 in the United States and Central America.

96   *everything from the neck up:* Lorraine Brennan, interview, November 14, 2002.

96   *acrylic cylinders:* The HBO chambers at Mass Eye and Ear are Secrist monoplace chambers, which fully enclose and isolate single patients. Some facilities have HBO chambers that can accommodate four to six patients at a time. These so-called multiplace chambers are small rooms where nurses, respiratory therapists, or technicians can tend to the patients, making it a bit easier to deal with some issues. The entire space is pressurized; the patients get oxygen through a mask, endotracheal tube, or hood, while the staff breathes room air under pressure. Once the door shuts, everyone is stuck inside until the session is over and the pressure is normalized. It does not seem to be the best use of a nurse, respiratory therapist, or technician's time.

100   *It interferes with oxygen delivery . . . no neurological effects:* DL Traber, DN Herndon, K Soejima, "The Pathophysiology of Inhalation Injury," in *TBC*, chap. 16.

101 *The half-life of carbon monoxide . . . less than half an hour:* This timing is approximate and varies with the individual and the COHb level. Different sources quote different estimates. The source I relied on is JC Fitzpatrick and WG Cioffi, "Diagnosis and Treatment of Inhalation Injury," in *TBC*, chap. 17.

102 *Practice Guidelines:* "Initial Management of Carbon Monoxide and Cyanide Exposure," in PGBC, chap. 4.

102 *"It was unambiguous":* And throughout chapter, John Schulz, interview, November 15, 2002.

103 *Bob Droste is no fan:* Bob Droste, interview, October 7, 2002.

104 *it was hardest on . . . Jack and Peggy:* The O'Sheas' thoughts, feelings, and experiences during Dan's time on Bigelow 13, interview, September 18, 2002.

105 *He makes no bones:* And throughout chapter, Mike Wilson, interview, June 20, 2002.

106 *A study Mike and Colleen Ryan did:* MD Wilson and CM Ryan, "A Program for the Integration of Spirituality into the Treatment of Burns," poster 217, annual meeting of the American Burn Association, 2001.

Chapter 7: At the Bedside

109 *"The trained nurse":* W Osler, "Nurse and Patient," in *Aequanimitas, with Other Addresses to Medical Students, Nurses, and Practitioners of Medicine*, 2d ed. (Philadelphia: P. Blakiston's, 1925).

109 *watch the morning routine:* The activities and interactions described in this chapter are based on observations made over the course of numerous visits to the Sumner Redstone Burn Center between March 2002 and April 2003. As is true throughout the book, the nurses, doctors, and other staff members are real and their names, when used, are their real names. The specific nursing activities described did not necessarily all take place on the same day, though they represent typical routines repeated on a daily basis.

110 *"In order to really give good care":* Mary Williams, interview, November 15, 2002.

110 *"Usually, we're more involved":* Nancy Parent's recollection, interview, October 27, 2002.

111 *close to 40,000 yards:* Estimate from Carolyn Washington, operations coordinator for Bigelow 13, telephone interview, November 19, 2002.

111 *huge gauze sandwiches:* These large, ready-made absorptive dressings have replaced the ones that were once prepared by hand by layering balls of "resilient fluff" between sheets of absorbent cotton and an outer layer of

gauze, then folding and autoclaving them. Making up these dressings in a range of shapes and sizes to fit various parts of the body was enormously time-consuming. The technique, developed in the 1950s at the U.S. Army Institute of Surgical Research, was considered a time-saving improvement over individually wrapped dressings. After detailing the method and materials (dressings to fit an adult arm and chest required fifty-four balls of fluff, a leg required ninety), the text observed, "A group of 10 volunteers can construct approximately 50 dressings in a two-hour period." "Dressings," in *Burns: A Team Approach,* ed. CP Artz, JA Moncrief, BA Pruitt (Philadelphia: Saunders, 1979), chap. 3.

114 *"if I went to hell":* And throughout chapter, Sally Morton, interview, June 21, 2002.

115 *Tom Parent still remembers his first morning:* Tom and Nancy Parent, interview, October 27, 2002.

116 *"I can't micromanage":* And throughout chapter, Tony DiGiovine, interview, July 19, 2002.

116 *"I'm Tony's partner":* Carolyn Washington, interview, October 10, 2002.

116 *the General's euphemistic titles:* The word *associate* is ubiquitous at the General. In addition to the unit service associates and operations associates, there are the patient care associates, formerly known as nurse's aides. The attendings call their professional corporation Burn Associates, and the entire hospital community is under the umbrella of Partners Health Care. It is an open question whether anyone buys this linguistic warmth and fuzziness and whether it makes MGH a better place for patients and employees. On Bigelow 13 they call the patient care associates burn techs, and at least one of the housekeepers refers to herself as a "USA."

117 *The General has a problem:* Blacks are not the only minority group in Boston or at the General, but they are the second largest racial group in the city, according to the 2000 U.S. census. A little more than 320,000 of Boston's 589,141 residents are white; 149,000 identify themselves as black or African American. Latinos (white and nonwhite) number about 85,000 and Asians about 44,000, which is fewer than the 46,000 recorded by the census as "some other race." Many people who work at the General, as well as patients, live outside the city, in the tony suburbs and working-class towns that make up the greater Boston area. Still, members of minority groups at the bedside are far outnumbered by those in the beds. Table DP–1, "Profile of General Demographic Characteristics: 2000, Geographic Area: Boston City, Massachusetts," U.S. Bureau of the Census, Census 2000.

117 *"to facilitate and promote":* Mission statement, Multicultural Affairs Office, Massachusetts General Hospital.

118 *The American Hospital Association estimates:* "The Nursing Shortage: Causes, Impact and Innovative Remedies" (testimony of the American Hospital Association before the United State House of Representatives Committee on Education and the Workforce, September 25, 2001).

118 *Nursing school enrollment:* Table 404: "Number of Basic RN Programs, Students and Graduates, Selected Academic Years 1950–51 to 1996–97," *U.S. Health Workforce Personnel Factbook,* National Center for Health Workforce Information and Analysis, Bureau of Health Professions, Health Resources and Services Administration, U.S. Department of Health and Human Services.

Testifying before a U.S. Senate subcommittee in 2001 on behalf of the American Nurses Association, Kathryn Hall, MS, RN, CNAA, painted a gloomy picture of an aging and shrinking nursing workforce caring for an aging and swelling patient population. "The lack of young people entering nursing has resulted in a steady increase in the average age of nurses. . . . Nationally, the average working RN is over 43 years old. The national average is projected to continue to increase before peaking at age 45.5 in 2010. At that time, large numbers of nurses are expected to retire and the total number of nurses in America will begin a steady decline. At the same time, the need for complex nursing services will only increase" (testimony of the American Nurses Association on the Nursing Shortage and Its Impact on America's Health Care Delivery System before the Subcommittee on Aging Committee on Health, Education, Labor, and Pensions, February 13, 2001).

120 *Advanced practice nursing:* It is also seen as an answer to the shortage of general practice and primary care physicians, especially in medically underserved areas. Advanced practice nurses can deliver primary care at lower cost than MDs and, since 1998, Medicare and Medicaid will reimburse for their services, thanks to an amendment of the 1997 Balanced Budget Act governing the Health Care Financing Administration (now called the Centers for Medicare and Medicaid Services, U.S. Department of Health and Human Services), which removed restrictions on practice settings, an action that caused an uproar in the medical community and an unsuccessful campaign by the American Medical Association to keep APNs under the supervision of MDs.

120 *"Nurses are a hard sell":* And throughout chapter, Mary-Liz Bilodeau, interview, March 6, 2002.

122 *During one of the informal break time chats:* Burn ICU, June 21, 2002.

123 *"I think it's personality":* Jennifer Verbesey, telephone interview, March 2, 2003.

125 *"There are thirty other women":* Frank Ireland, interview, March 6, 2002.

125 *men representing about 5 percent:* Men tend to be steered toward EMT/paramedic careers rather than nursing, but the good news is an apparent upward trend in the numbers of men entering nursing, thanks in part to aggressive campaigns to give the profession a macho edge. In 1984, men represented 3.3 percent of all RNs; it had risen to 5.4 by 1996. Table 405: "Registered Nurses Employed in Nursing by Gender and Race/Ethnicity, Selected Years, 1984–1996," in *U.S. Health Workforce Personnel Factbook,* National Center for Health Workforce Information and Analysis, Bureau of Health Professions, Health Resources and Services Administration, U.S. Department of Health and Human Services.

125 *"One day when I was banging nails":* Rick MacDonald, interview, June 20, 2002.

Chapter 8: Gasping for Breath

130 *"To be totally frank":* Nancy Giese, interview, May 2, 2003.

131 *"everybody comes to your rescue":* Nancy Giese, interview, June 21, 2002.

131 *The best example of the teamwork:* Nancy Giese, Bob Droste, and Mike Wilson took turns telling what happens when a new admit arrives; interview, October 7, 2002.

133 *she is already at work:* Observation, June 21, 2002.

135 *400 times the heat-carrying capacity of air:* FC DiVincenti, BA Pruitt, JM Reckler, "Inhalation Injuries," *J Trauma* 11 (1971): 109–117.

135 *a circus fire-eater:* Pauline Buskiewicz, interview, March 6, 2002.

136 *"little branching trees":* John Schulz, interview, November 15, 2002.

136 *a claim Mary-Liz Bilodeau disputes:* Written communication, October 22, 2003; telephone interview, October 28, 2003. "The bed company will tell you this works, but it really doesn't."

137 *"the hallmark of his admission":* And throughout chapter, John Schulz, interview, October 10 and November 15, 2002.

138 *"being buried alive":* And throughout chapter, Mike Wilson, interview, October 7, 2002.

140 *he was horrified:* Unless otherwise noted, throughout chapter, Mike Marley, interview, October 7, 2002.

142 *"The residents get a couple of weeks":* Mike Marley, e-mail, July 14, 2003.

143 *one man they remember:* And throughout chapter, Jack and Peggy O'Shea's recollections, observations, and feelings, interview, September 18, 2002.

144 *"They were great parents":* Mary Williams, interview, November 15, 2002.

Chapter 9: The Scourge and the Silver Standard

145 *Nancy Parent was exhausted:* And throughout chapter, Tom and Nancy Parent's words, thoughts, and feelings, interview, October 27, 2002.

147 *"It's exhausting":* Mike Tiffany, interview, June 21, 2002.

148 *Ashley said to her:* Recollected by Nancy Parent, interview, October 27, 2002.

151 *"sinking into a hectic state":* Quoted in Artz, Moncrief, Pruit, *Burns*, chap. 1.

151 *James Lister introduced the concept of antisepsis:* Interestingly, the first free-standing burn unit in the world was established in Edinburgh in 1848 by James Syme, Lister's father-in-law. Syme's great innovation was to isolate burn patients in an entirely separate building at a time when hospitals were, quite literally, hotbeds of infection. Cockshott, "History of the Treatment of Burns." Twenty years after his father-in-law opened his burn hospital, Lister published *On the Antiseptic Principle of the Practice of Surgery* (1867), advocating the use of carbolic acid and heat-sterilized instruments in the operating room, a practice that revolutionized surgery.

151 *between 1966 and 1975: TBC,* introduction.

151 *"Prior to effective topical therapy":* J Moncrief, "The Development of Topical Therapy" [American Burn Association presidential address, 1971], *J Trauma* 11 (1971): 906–910.

152 *As John Schulz explained it:* E-mail, February 6, 2003.

153 *"You put some slop on the burn":* John Burke, interview, April 9, 2002.

153 *James Earle observed . . . preparations for treating burns:* HJ Klasen, "A Historical Review of the Use of Silver in the Treatment of Burns, I. Early Uses," *Burns* 26 (2000): 117–130. Throughout this chapter I rely on Klasen's fascinating article for much of the historical information about silver.

154 *"There's always new dressing materials":* Mary-Liz Bilodeau, interview, May 5, 2003.

154 *since classical times:* Klasen, "Historical Review."

154 *"If you prevent infection":* Paracelsus, quoted in WW Monafo, "Past Is Prologue: Two Theo's and a Friend" [American Burn Association presidential address, 1977], *J Trauma* 17 (1977): 785–792.

155 *a dangerously strong 10 percent solution:* Clowes, Lund, and Levenson outlined the U.S. Army protocol in their paper, "The Surface Treatment of Burns" (*Ann Surg* 118 [1943]: 761–779), which prescribed one application of 10 percent tannic acid, followed by four applications (at half hour intervals) of a solution combining 5 percent tannic acid and 5 percent silver nitrate. By today's standards, it was a staggering chemical assault on a wound.

156 *cannot be used internally:* Ingesting silver causes a condition called argyria, which gives the skin a bluish cast, undoubtedly the origin of the term *blue blood*, used to describe aristocrats "born with a silver spoon in their mouths." Aside from looking odd, this condition is not harmful, though it is irreversible.

156 *Moyer and his colleagues:* CA Moyer, L Brentano, DL Gravens, HW Margraf, WW Monafo, "Treatment of Large Human Burns with 0.5% Silver Nitrate Solution," *Arch Surg* 90 (1965): 812–867.

157 *Acticoat is basically:* There are a few other silver powder–impregnated dressings available or in development, but Acticoat is by far the most widely used. Perhaps that is because of its clear superiority, or because it was the first out of the box and has the marketing muscle of pharmaceutical giant Smith & Nephew behind it.

158 *turns everything . . . brownish black:* Silver nitrate solution is a colorless, odorless liquid, but it darkens when exposed to air, an oxidation process that results when the silver ion grabs oxygen to become silver oxide. It is the same chemical change that makes black-and-white photography possible.

158 *John Burke, the mastermind:* Interview, June 19, 2002.

159 *"Our burn techs are wonderful":* Nancy Giese, interview, June 21, 2002.

160 *she has an excellent working relationship:* And throughout chapter, Edna Gavin, interview, June 20, 2002.

160 *"It's very difficult":* And throughout chapter, Carl Baxter, interview, November 14, 2002.

162 *"It felt like a wire brush":* "The Burn Survivor's Perspective," plenary session II, annual meeting of the American Burn Association, April 20, 2001.

162 *"Most people try":* James Burke, interview, June 19, 2002.

162 *"I think it's a terrible idea":* And throughout chapter, Bob Droste, interview, May 5, 2003.

165 *John Burke's capsule history:* Interview, April 9, 2002.

Chapter 10: Cutting Off the Dead

169 *"I got pissed off":* And throughout chapter, Terri Leddy, interview, October 29, 2002. Information about how the OR floor and the burn OR work is drawn from this interview, as well as interviews with John Schulz and Jennifer Verbesey, and observations during two long days in the burn OR, in June and July 2002. Details of the procedures and patient histories are composites to protect patient privacy as well as to represent a broad range of the OR experience.

171 *John Findley . . . told me:* Interview, July 16, 2002.

173 *the idea can be traced:* O Cope, JL Langohr, FD Moore, RC Webster, "Expeditious Care of Full-Thickness Burn Wounds by Surgical Excision and Grafting," *Ann Surg* 125 (1947): 1–22.

173 *"pain and suffering":* MJ Muller, D Ralston, DN Herndon, "Operative Wound Management," in *TBC*, chap. 12.

173 *Burke began thinking about the problem:* And throughout chapter, John Burke, interview, June 19, 2002.

173 *Sir Ashley Miles:* The eminent microbiologist Sir Arnold Ashley Miles (1904–1988) was director of the Lister Institute of Preventive Medicine from 1952 to 1972. His landmark studies of bacterial infections in combat wounds during World War II pinpointed the problem of cross-contamination and led to improved infection-control practices in the surgical setting.

175 *a paper espousing early excision:* Z Janzekovic, "A New Concept in the Early Excision and Immediate Grafting of Burns," *J Trauma* 10 (1970): 1103–1108.

175 *was able to document:* JF Burke, CC Bondoc, WC Quinby, "Primary Burn Excision and Immediate Grafting: A Method for Shortening Illness," *J Trauma* 14 (1974): 389–395. The study followed 200 burned children treated at the Boston Shriners Hospital for Children, where Burke was chief of staff and his coauthors were attending surgeons, though then, as now, the MGH and Shrine surgical staff overlapped. Half of the patients had their burns treated with silver nitrate, but grafting was delayed until eschar had separated; half had their burns excised and grafted, with both donor and graft sites treated with silver nitrate. The study found that "excision and immediate grafting, combined with silver nitrate, reduced the time of burn illness between one-third and one-half of the time required for a similar sized injury to recover if treated with silver nitrate therapy alone."

176 *"The earlier one can excise":* Burke, Bondoc, Quinby, "Primary Burn Excision."

176 *The result was a precipitous decline:* RG Tompkins, JF Burke, DA Schoenfeld et al., "Prompt Eschar Excision: A Treatment System Contributing to Reduced Burn Mortality," *Ann Surg* 204 (1984): 272–281.

177 *"If it needs grafting":* John Schulz, interview, November 15, 2002.

178 *"don't operate on Tuesdays":* John Schulz, interview, July 17, 2002. The Monday, Wednesday, Friday schedule is imposed by the OR administration, but exceptions are made when circumstances demand it. Life-saving surgery takes place when it has to. After the Rhode Island fire, Schulz and his associates operated seven days a week, most of the time with two rooms going. In the first week alone, each of the eleven patients at the General made a minimum of two trips to the OR and continued to have surgery in stages for many weeks afterward.

182 *used on Clifford Johnson:* Johnson's story is a favorite of Cocoanut Grove historians, not only because of his own heroism and that of the doctors and nurses who treated him, nor just because he was the first known to survive such massive burns, but also because of the way he ended up dying. After a one-year stay at Boston City Hospital and another of rehabilitation in several military facilities, he went home to Missouri, taking with him as his wife one of the BCH nurses who had cared for him. A scant decade later, a jeep he was driving went off the road and burst into flames. Trapped inside the burning vehicle, Johnson finally met his death by fire. Details about his age at the time of the fire, the size of his burns, the length of his hospital stay, and the circumstances of his fatal accident differ, depending on who is telling the story. This version, including the Gillette blue blade, comes from the recollections of Stanley Levenson, telephone interviews, October 14, 2001, and July 2, 2002; video.

185 *"All kinds of people":* This sort of traffic may provide comic relief, but it is generally regarded as a nuisance. Some time after this particular visit, the OR administration at MGH implemented new policies to limit the presence of nonessential personnel in the operating rooms to those with prior and approved appointments.

Chapter 11: Healing

191 *Tom Parent's left arm:* And throughout chapter, Tom and Nancy Parent's recollections, interview, October 27, 2002.

193 *"Paradoxically, the same systemic":* HA Linares, "Pathophysiology of the Burn Scar," in *TBC*, chap. 44.

194 *a host of growth factors:* Their names hint at where they come from and the roles they play: epidermal growth factor (EGF), fibroblast growth factor (FGF), keratinocyte growth factor (KGF), platelet-derived growth factor (PDGF), transforming growth factor (TGF), vascular endothelial growth factor (VEGF), to name just a few. But growth factors are far more complicated than their names imply. Many are secreted by more than one type of cell and have more than one role at various points along the way.

194 *"same cellular and molecular structures":* Linares, "Pathophysiology of the Burn Scar."

194 *The complexity of the interaction:* DG Greenhalgh, "Wound Healing," in *TBC*, chap. 43; M Spies, MJ Muller, DN Herndon, "Modulation of the Hypermetabolic Response After Burn," in *TBC*, chap. 27. Regranex, a topical gel containing platelet-derived growth factor, has been approved by the FDA for diabetic foot ulcers, the classic nonhealing wound, and recombinant hu-

man growth hormone has been shown to improve donor site healing in children. DN Herndon, RE Barrow, KR Kunkel et al., "Effects of Human Growth Hormone on Donor-Site Healing in Severely Burned Children," *Ann Surg* 212 (1990): 424–432. Another avenue of research is gene therapy—splicing growth factor genes onto keratinocytes or fibroblasts to induce them to secrete the growth factors at the optimal time and place. Nothing yet qualifies as a silver bullet, but research in this area continues.

199 *invasive form of skin cancer:* This consequence of a poorly healed burn is known as Marjolin's ulcer, named for Jean Nicolas Marjolin, the nineteenth-century French surgeon who described cancers arising from deteriorating burn scars, in his *Dictionnaire de mèdecine pratique* (1828).

200 *an overabundance of transforming growth factor-beta 1:* PG Scott, A Ghahary, EE Tredget, "Molecular and Cellular Basis of Hypertrophic Scarring," in *TBC*, chap. 43; DA Bettinger, DR Yager, RF Diegelmann, IK Cohen, "The Effect of TGF-ß on Keloid Fibroblast Proliferation and Collagen Synthesis," *Plast Reconstr Surg* 98 (1995): 827–833; A Ghahary, YJ Shen, PG Scott, Y Gong, EE Tredget, "Enhanced Expression of mRNA for Transforming Growth Factor-beta, Type I and Type III Procollagen in Human Post-Burn Hypertrophic Scar Tissues," *J Lab Clin Med* 122 (1993): 465–473; R Wang, A Ghahary, Q Shen, PG Scott, K Roy, EE Tredget, "Hypertrophic Scar Tissues and Fibroblasts Produce More Transforming Growth Factor-ß1 mRNA and Protein Than Normal Skin and Cells," *Wound Rep Reg* 8 (2002): 128–137.

202 *Others argue:* One of the most vocal skeptics of compression therapy is G. Patrick Kealey, MD, director of the University of Iowa Burn Center, who has conducted clinical trials comparing custom-made and off-the-shelf pressure garments, which were found to have equal effect (GP Kealey et al., "Prospective Randomized Comparison of Two Types of Pressure Therapy Garments," *J Burn Care Rehab* 11 [1990] :334–336), and more damningly, pressure garments versus nothing, also equally effective (P Chang, KN Laubenthal, RW Lewis, 2d, MD Rosenquist, P Lindley-Smith, GP Kealey, "Prospective, Randomized Study of the Efficacy of Pressure Garment Therapy in Patients with Burns," *J Burn Care Rehab* 16 [1995]: 473–475). On the other side of the argument, electron microscopy has shown the alteration in both subsurface collagen arrangement and surface contours before and after pressure over a period of months. An example can be found in HA Linares, "Pathophysiology of the Burn Scar," in *TBC*, chap. 44, fig. 44.5.

202 *Some . . . concede it is wishful thinking:* Discussion, "Pressure Garments: Yes or No?" (luncheon symposium, annual meeting of the American Burn

Association, April 26, 2002; Patrick G. Kealey, MD, and Roberta Mann, RN, moderators). One participant voiced a common sentiment: "I don't use them very often and when I do it's really not for scar control, but for psychological support. Some people think they need pressure garments and so we give them pressure garments to make them feel better."

203 *"For a while I was wearing them"*: And throughout chapter, Dan O'Shea, interview, August 18, 2002.

204 *more than half*: Mary-Liz Bilodeau, telephone interview, October 28, 2003.

204 *"Long-term?"*: John Schulz, interview, October 10, 2002.

205 *every day reminded them*: And throughout chapter, Jack and Peggy O'Shea interview, September 18, 2002.

205 *"It's not the same"*: And throughout chapter, interview, Nancy Giese, May 2, 2003.

206 *"They fade back"*: And throughout chapter, Bob Droste, interview, May 5, 2003.

206 *"Emergence from sedation"*: And through end of chapter, John Schulz, interview, November 15, 2002.

207 *"Dan's parents really struggled"*: And throughout chapter, Mike Wilson, interview, October 7, 2002.

207 *"people say to me"*: John Findley, interview, July 16, 2002.

208 *"patients begin to realize"*: Carla Cucinatti, interview, June 20, 2002.

Chapter 12: New Skin

213 *Tom and Nancy do not agree*: Tom and Nancy Parent, interview, October 27, 2002.

214 *some colorfully grotesque attempts*: Artz, Moncrief, Pruitt, "Operative Care: Grafting," in *Burns*, chap. 4.

214 *Stanley Levenson told me a story*: Telephone interview, October 28, 2003.

215 *"You'd be amazed"*: John Schulz, e-mail communication, November 1, 2003.

216 *Biobrane*: Manufactured by Bertek Pharmaceuticals; approved by the FDA in 1983.

217 *TransCyte*: Developed by Advanced Tissue Sciences, which is in receivership, and subsequently acquired by Smith & Nephew, it was initially approved and marketed under the name Dermagraft-TC. The name change was approved by the FDA in 1998, as was its use under that name as a temporary wound covering for burn wounds that are expected to heal without grafting and deeper ones awaiting grafting. www.fda.gov/cdrh/pma/pmaaug98.html; PMA Final Decisions Rendered for August 1998:

supplemental approvals; U.S. Food and Drug Administration, Center for Devices and Radiological Health.

218 *the linchpin of early excision:* Burke, Bondoc, Quinby, "Primary Burn Excision and Immediate Grafting."

218 *Integra:* Originally licensed to Marion Laboratories; Johnson & Johnson now holds the marketing rights.

218 *"to get into the modern world":* And throughout chapter, John Burke, interview, June 19, 2002.

220 *Yanni Yannas:* Ioannis V. Yannas, Ph.D., now a professor in the Department of Mechanical Engineering and the Division of Bioengineering and Environmental Health, both at MIT, and in the joint Harvard-MIT Program in Health Sciences and Technology.

221 *a high art:* One of those places is the University of Washington Burn Center at Harborview in Seattle. Dr. David Heimbach, the former medical director at Harborview, was once introduced to a gathering as the man who uses Integra for everything except wrapping peanut butter and jelly sandwiches (Johnson & Johnson–sponsored dinner, annual meeting of the American Burn Association, April 19, 2001).

A joint report from Harborview and the surgery department at Tokyo Medical University offered advice on how to get a better result by grafting an ultrathin split-thickness graft rather than the extremely fragile epidermal graft recommended by the manufacturer. P Fang, LH Engrav, NS Gibran et al., "Dermatome Setting for Autografts to Cover Integra," *J Burn Care Rehabil* 23 (2002): 327–332. DM Heimbach, GD Warden, A Luterman et al., "Multicenter Postapproval Clinical Trial of Integra Dermal Regeneration Template for Burn Treatment," *J Burn Care Rehabil* 24 (2003): 42–48, reported the results of a study of infection rates requested by the FDA. It described infection prevention strategies devised by the thirteen burn care facilities that participated in the trial, including meshing of Integra, the use of a range of topical antimicrobial and dressings, frequent wound inspection, and timely treatment of infection in wounds not covered with Integra.

221 *"Integra is not a miracle":* John Schulz, interview, November 15, 2002. Nonetheless, Schulz and his colleagues at MGH published a study documenting that Integra decreased the amount of time severely burned patients spent in the hospital, which significantly lowers the total cost of treatment in addition to benefiting the patient's quality of life. CM Ryan, DA Schoenfeld, M Malloy, JT Schulz, RL Sheridan, RG Tompkins, "Use of Integra Artificial Skin Is Associated with Decreased Length of Stay for Severely Injured Adult Burn Survivors," *J Burn Care Rehabil* 23 (2002): 311–317.

222 *first paper on the process:* JG Rheinwald, H Green, "Serial Cultivation of Strains of Human Epidermal Keratinocytes: The Formation of Keratinizing Colonies from Single Cells," *Cell* 6, (1975): 331–343. This was followed by H Green, O Kehinde, J Thomas, "Growth of Cultured Human Epidermal Cells into Multiple Epithelia Suitable for Grafting," *Proc Natl Acad Sci* 76 (1979): 5665–5668.

222 *"wet tissue paper":* MJ Muller, D Ralston, N Herndon, "Operative Wound Management," in *TBC,* chap. 12.

222 *According to Genzyme:* www.genzymebiosurgery.com.

223 *as close as you can get to embryonic stem cells:* Closer yet is amniotic membrane, which is used as a temporary biological wound dressing in some parts of the world, though not in the United States. This very fragile tissue has to be replaced daily, as it tends to disintegrate. Moreover, it carries a high risk of viral and bacterial contamination. Synthetic and biosynthetic dressings serve the same purpose and are more durable and safer from the point of view of disease transmission.

223 *One product that is still in development:* ST Boyce, RJ Kagan, NA Meyer, KP Yakuboff, GD Warden, "The 1999 Clinical Research Award: Cultured Skin Substitutes Combined with Integra Artificial Skin to Replace Native Skin Autograft and Allograft for the Closure of Excised Full-Thickness Burns," *J Burn Care Rehabil* 20 (1999): 453–461; and comments, *J Burn Care Rehabil* 22 (2001): 197–199.

223 *the FDA Center for Devices and Radiological Health:* The exception is Alloderm, which is regulated as human tissue for transplantation by the FDA Center for Biologics Evaluation and Research, CFR Title 21, Part 1270. www.fda.gov/cber/tiss.htm. Although Epicel would seem to fall into this category, Genzyme opted for a humanitarian device exemption (HDE), which was granted in 1997, presumably because it was a less cumbersome undertaking for a product that is hardly a runaway best seller.

224 *Alloderm is cadaver skin:* Alloderm is manufactured by LifeCell Corporation, and is approved as human tissue for transplantation for use in burns. www.fda.gov/cber/tissue/regotab.pdf; Proposed New Regulatory Framework for Human Tissue.

225 *Alloderm must be screened:* In 2001, an eighteen-piece lot of Alloderm had to be recalled after distribution when the donor tested positive for HIV. LifeCell insists viruses are rendered nonviable during the manufacturing process and thus Alloderm cannot transmit disease. There is no indication in the recall notice that HIV was detected in the product, only that the tissue from which it was made had been "collected from a donor in which HIV–1 PCR testing detected HIV DNA." FDA Enforcement Report; Re-

calls and Field Corrections: Biologics, September 5, 2001; www.fda.gov/
bbs/topics/ENFORCE/2001/ENF00709.htm.

225 *Dermagraft is a synthetic:* Dermagraft was developed by Advanced Tissue
Sciences and acquired by Smith & Nephew. It was approved in 2001 for
diabetic foot ulcers; www.fda.gov/cdrh/pdf/p000036a.pdf. In 2003, under
a humanitarian device exemption, it was approved for the treatment of epidermolysis bullosa, a rare and terrible skin disease that has nothing to do
with burns. www.fda.gov/cdrh/ode/hdeinfo.html.

225 *Apligraf and OrCel are sandwiches:* OrCel is made by Ortec International
and was approved for burn donor sites in 2001. www.fda.gov/cdrh/
pma/pmaaug01.html; PMA Final Decisions Rendered for August 2001:
original approvals; U.S. Food and Drug Administration, Center for Devices and Radiological Health. It was also approved that same year as a humanitarian use device for epidermolysis bullosa, and is in the clinical trial
stage for diabetic and venous ulcers. www.fda.gov/cdrh/ode/hdeinfo.html;
Humanitarian Use Devices; U.S. Food and Drug Administration, Center
for Devices and Radiological Health.

Apligraf is approved for venous leg ulcers and diabetic foot ulcers. It
applied for and received multiple FDA approvals, the first in 1998 and the
most recent, for introducing a new cell strain into its production, in 2002.
www.fda.gov/cdrh/pma/pmajul02.html; PMA Final Decisions Rendered
for May 1998: original approvals; July 2002: supplemental approvals; U.S.
Food and Drug Administration, Center for Devices and Radiological
Health. It has been a bit of a football, kicked back and forth between
Organogenesis, the biotech company that developed it, and the pharmaceutical giant Novartis. For a while, it was unavailable as the two companies wrangled over marketing rights. Novartis ultimately returned the
rights to Organogenesis, which operated for a time under Chapter 11
bankruptcy protection. As of this writing, Organogenesis has emerged
from bankruptcy and has resumed both manufacture and distribution of
Apligraf. In the world of burn treatment, this has made little difference because Apligraf has not gained wide acceptance. "To my mind it has no application distinct from cadaveric skin, allograft," John Schulz told me
dismissively. Interview, November 15, 2002.

226 *"Any port in a storm":* John Schulz, interview, November 15, 2002.

Chapter 13: Burned to Death

227 *"Sometimes things just happen":* And throughout chapter, Nancy Giese, interview, May 2, 2003.

228 *"We have very few code calls here":* And throughout chapter, Bob Droste, interview, May 5, 2003.

230 *Consensus Development Conference on Supportive Therapy in Burn Care:* The conference, held on November 10 and 11, 1978, was a joint effort of the American Burn Association and National Institute of General Medical Sciences, one of the National Institutes of Health.

230 *maximal therapy is imperative:* BE Zawacki, "The Doctor–Patient Covenant," in NI General Medical Sciences with ABA, "Consensus Development Conference on Supportive Therapy in Burn Care," *J Trauma* 19, suppl (1979): 855–936.

230 *"moral responsibility to initiate therapy":* "Consensus on Supportive Therapy."

230 *over issues of life and death:* BE Zawacki, "Tongue-Tied in the Burn Intensive Care Unit," *Critical Care Medicine* 17 (1989): 198–199. Zawacki, who retired as director of the Burn Center at Los Angeles County/University of Southern California Medical Center, and is associate professor emeritus at USC in both the School of Medicine and the School of Religion, has authored or coauthored numerous articles on ethical decision-making in medicine in general and specifically in burn care, including: "The 'Futility Debate' and the Management of Gordian Knots," *J Clin Ethics* 6 (1995): 112–127; "Ethical and Practical Considerations in Decision Making in Burn Care," *Top Emerg Med* 3 (1981): 1–5; SH Imbus, BE Zawacki, "Encouraging Dialogue and Autonomy in the Burn Intensive Care Unit," *Crit Care Clin* 2 (1986): 53–60; SH Imbus, BE Zawacki, "Autonomy for Burned Patients When Survival Is Unprecedented," *N Engl J Med* 297 (1977): 308–311; J Starr, BE Zawacki, "Voices from the Silent World of Doctor and Patient," *Camb Q Healthc Ethics* 8 (1999): 129–138.

231 *"I believe that in a process":* Mike Wilson, interview, June 20, 2002.

231 *"We do our best":* John Schulz, interview, November 15, 2002; written communication, October 22, 2003.

233 *Mary-Liz Bilodeau remembers:* And throughout chapter, interview, June 20, 2002.

234 *an article . . . that proposed:* CM Ryan, DA Schoenfeld, WP Thorpe, RL Sheridan, EH Cassem, RG Tompkins, "Objective Estimates of the Probability of Death from Burn Injuries," *N Engl J Med* 338 (1998): 362–366.

234 *Peggy's response was:* Peggy O'Shea, interview, September 18, 2002.

235 *his own filter:* Dan O'Shea, interview, August 18, 2002.

235 *"We're here to make it unsad":* John Schulz, interview, July 17, 2002.

235 *"bad save":* Susan DiBattista, interview, July 16, 2002.

236 *shook Peggy and Jack:* Peggy and Jack's thoughts, feelings, and experiences, interview September 18, 2002.

236 *"We're not easy to demoralize":* John Schulz, interview, June 20, 2002.

237 *John Findley disagrees:* Interview, July 16, 2002.

237 *"It's very emotionally taxing":* John Schulz, interview, November 15, 2002.

Chapter 14: A New Life

239 *Dan O'Shea sat on a bench:* The thoughts, feelings, and words attributed to Dan O'Shea and his parents, as well as the events described, interviews August 18, 2002 (Dan) and September 18, 2002 (Peggy and Jack).

245 *They might not have let:* The story of Tom Parent's discharge and adjustment in the first year after he was burned, the words spoken by Tom and Nancy, and those attributed to others by them, interview, October 27, 2002.

247 *The outpatient clinic:* Interviews, June 20, 2002, and May 5, 2003.

248 *She spied the mother:* Interview, May 5, 2003.

248 *"Taking care of patients":* Mary-Liz Bilodeau, interview, May 5, 2003.

249 *John Findley has observed:* Interview, July 16, 2002.

249 *John Burke is convinced:* Interview, June 19, 2002.

251 *"I wanted to be able to intervene":* Mary-Liz Bilodeau, written communication, October 22, 2003.

254 *"It's a blessing":* Mike Wilson, interview, October 7, 2002.

Acknowledgments

There are many people without whose help and insights I could not have written this book. I must begin by acknowledging Stanley Levenson, MD, who was there from the beginning and gave me the same gentle but persistent prodding that he must have given my father when they roomed together at Harvard Medical School more than sixty years ago.

I am boundlessly grateful for the red carpet extended to me at Massachusetts General Hospital, especially its Sumner Redstone Burn Center. But the MGH burn unit is only one of many in the country where patients are tended by burn care professionals every bit as dedicated as the ones I introduce in these pages. For reasons of narrative economy, I shine the spotlight on a handful of the people working in this particular and special place and tell only a few of the stories of the lives they have touched; I wish I could have named them all and told every story. In lieu of that, these pages contain my thanks and tribute to them.

At Mass General: nurse manager Tony DiGiovine and nurses Lynn Bellavia, Peter Dawber, Susan DiBattista, Bob Droste, Nancy Giese, Frank Ireland, Leslie Loui, Rick MacDonald, Betsy Merriam, Sally Morton, Judith Underwood-DerAnanian, Mary Williams, and Mike Wilson, all on Bigelow 13, and Terri Leddy in the OR and Kelley Burke in the Emergency Department; operations coordinator Carolyn Washington, operations associate Darlene Alcinder, unit service associate Marian Henry, patient care associates Carl Baxter and Edna Gavin; case manager Pauline Buskiewicz; rehabilitation therapists Stacey Connelly, OT, Laura Kastrenos, PT, Mike Marley, RT, and Mike Tiffany, PT; Carla Cucinatti, MSW, LCSW; ethicist Christine Mitchell, RN; Sharon M. Burke, patient care coordinator for the MGH Burn Associates, and Pat Kelleher, administrative manager of Trauma and Burn Services; John Findley, MD;

residents David T. Cooke, MD, and Jennifer Verbesey, MD, who have since moved on in their medical careers; burn center codirectors Colleen M. Ryan, MD, and Robert L. Sheridan, MD, whose fleeting presence in this book does not adequately reflect their central role in caring for burn patients; Sue McGreevey, manager of science and research communications, public affairs office; Katherine Flaherty, director of Medicaid and uncompensated care programs, Partners Health Care; and financial services counselor Cathy DeVitto. Particular thanks go to clinical nurse specialist Mary-Liz Bilodeau, John Schulz, MD, and the incomparable John F. Burke, MD, without all of whom this book would have been a slighter and considerably more error-riddled thing.

At the Norman Knight Hyperbaric Medicine Center, Massachusetts Eye and Ear Infirmary, hyperbaric nursing coordinator Lorraine Brennan, RN.

At the Shriners Hospital for Children (Boston), Janet Mulligan, RN, director of patient care services; Mary Jo Baryza, director of therapeutic services; nurse manager Alyson Roe, and nurses John Haugh, Karen Magoon, Krisha Sobocinski, Danielle Woods; Michelle Nankin, LICSW, director of family service and care coordination; Kathy Prelack, MS, RD; John Silva, general services department; and Daniel Driscoll, MD.

At New York-Presbyterian Hospital, Randolph Hearst Burn Center (Cornell Burn Unit), nurse manager Bob Dembicki, clinical nurse specialist Frank Costello, and nurses Eugene Ambrosio and Dan Haughie; Gary Conyers; and Palmer Bessey, MD, all of whom generously took time to give me my first glimpse of the world inside a burn unit.

Nicole S. Gibran, MD, director of the burn center at Harborview Medical Center, University of Washington, Seattle, deserves special thanks for introducing me to John Schulz.

I salute the openness, but most of all the courage, of the man who is called Dan O'Shea in this book, of his parents, Peggy and Jack O'Shea, and of Tom, Nancy, and Ashley Parent.

I am grateful for the help of Jeffrey Mifflin, archivist and curator of special collections, and for the rich holdings of the Treadwell Library at Massachusetts General Hospital; the National Library of Medicine and its wonderful PubMed search engine, an invaluable tool to a writer living in the boondocks; and the library at the New York Academy of Medicine, whose free and open access to the public makes it a particular treasure. Nor could I have written this book without the resources of the American Burn Association and the encyclopedic *Total Burn Care*, edited by David Herndon.

Gazing heavenward, I honor my patron saint, Berton Roueché, whose marvelous "Annals of Medicine" in the *New Yorker* I began to read at an early age, planting in me the idea that writing about medicine was a profession worth pursuing.

My gratitude goes to Peter Kelman and Therese Mageau, careful and critical early readers, for keeping me from going off the deep end; Judith Pucci, Kate Zentall, and Katie and Harry Sugarman, for their invaluable feedback and unflagging enthusiasm; Judy Motzkin and Sue Kelman, for their hospitality and good counsel; my agent, Lane Zachary, who did not think a book about burns was a totally crazy idea; and Marnie Cochran, my able and supportive editor, who believed her.

And finally, I thank my son, JC Ravage, who was considerate enough to grow up and go away to college so I could finally write this book.

Index

ABA (American Burn Association), *xv*, 49, 102, 252–253
    Practice Guidelines, 39, 102
Acticoat, 157, 158, 245
advanced trauma life support (ATLS), 50
African Americans on staff at Massachusetts General Hospital, 116–117
age of victim
    children, 30, 67–68, 69–70, 70, 201
    elderly, 30, 65–67
    and risk of death, 67, 233–234
    and risk of infection, 152
airway damage. *See* inhalation injury
alcoholism, 59, 61, 241–244
Aldrich, Robert Henry, 16
A-line (arterial line). *See* catheterization
Allen, Harvey Stuart, 17, 151
Alloderm, 224–225
allograft (cadaver skin), 214, 217–218
American Burn Association (ABA), *xv*, 49, 102, 252–253
    Practice Guidelines, 39, 102
American College of Surgeons Committee on Trauma (ASCOT), 49
American Hospital Association, 118
amputation, 29, 64, 66
anesthesiologists, 168
anger of burn victims, 240, 242–243, 251
animal skin grafts (xenografts), 213–214

antiseptic environments
    bacteria-controlled nursing unit, 79, 92–94, 111, 137–138
    operating room, 179
Apligraf, 225
arterial lines. *See* catheterization
artificial skin. *See* skin substitutes
ASCOT. *See* American College of Surgeons Committee on Trauma
ATLS (advanced trauma life support), 50
autografts, 214
autologous skin grafts. *See* autografts

bacteria
    antibiotic-resistant, 19, 150, 156, 157
    antimicrobial properties of silver, *xii*, 17, 108, 155–157
    on burn surface, 150
    clostridium, 19, 100, 150
    E. coli, 19, 150
    gram-negative, 150, 156
    gram-positive, 16, 150, 156
    in hospitals, 149
    klebsiella, 150
    methicillin-resistant *Staphylococcus aureus* (MRSA), 137–138
    pseudomonas, 150
    sources of, 152. *See also* infection
    staphylococcus, 16, 19, 150
    streptococcus, 19, 150
    tetanus, 54